Translating and Interpreting Specific Fields: Current Practices in Turkey

Ayşegül Angı (ed.)

Translating and Interpreting Specific Fields: Current Practices in Turkey

PETER LANG

Bibliographic Information published by the Deutsche Nationalbibliothek
The Deutsche Nationalbibliothek lists this publication in the Deutsche Nationalbibliografie; detailed bibliographic data is available in the internet at http://dnb.d-nb.de.

Library of Congress Cataloging-in-Publication Data
Names: Angı, Ayşegül editor.
Title: Translating and interpreting specific fields : current practices in Turkey / Ayşegül Angı [editor].
Description: Frankfurt am Main ; New York : Peter Lang, 2017.
Identifiers: LCCN 2017047410 | ISBN 9783631738504
Subjects: LCSH: Technology--Translating--Turkey. | Science--Translating--Turkey. | Translating and interpreting--Turkey. | Turkish language--Technical Turkish.
Classification: LCC T11.5 .T55 2017 | DDC 418/.036--dc23 LC record available at https://lccn.loc.gov/2017047410

Printed by CPI books GmbH, Leck.

ISBN 978-3-631-73850-4 (Print)
E-ISBN 978-3-631-74051-4 (E-PDF)
E-ISBN 978-3-631-74052-1 (EPUB)
E-ISBN 978-3-631-74053-8 (MOBI)
DOI 10.3726/b12893

Peter Lang – Frankfurt am Main · Bern · Bruxelles · New York · Oxford · Warszawa · Wien

This publication has been peer reviewed.

www.peterlang.com

Content

Assist. Prof. Ayşegül Angı Ph.D.

Marmara University
Department of Translation and Interpreting

Editorial

As a field of study[1] when Translation and Interpreting Studies (TIS) is examined from the interdisciplinary perspective, it interacts primarily with such fields as linguistics, literature, and sociology while constructing its theoretical framework. Its wide-ranging practices in communicating information and disseminating scientific and similar knowledge pertaining to the other (sub)fields of study such as engineering, international relations, law, medicine, and so on, contribute to the expansion of its theoretical realm as well. From the multidisciplinary and/or transdisciplinary point of view, the reciprocal relations with the other fields of study should be fostered in order to serve prospectively for the social ends[2]. However, when the translation and interpreting studies are retrospectively examined in terms of subject domains, despite the widespread practice regarding the other specific fields, a large number of the studies, surprisingly, focus on the literary translation[3]. Therefore, the increasing social and sectorial need for transdisciplinary training/research has instigated this preliminary process of gathering the individual studies in a book.

When specific fields are examined in terms of subject domains, it is essential to maintain a standardization since there have been a variety of classifications in practice. Obviously, it is much easier to categorize these domains broadly as

1 'Field of study' is defined by International Standard Classification of Education [(ISCED) developed by United Nations Educational, Scientific, and Cultural Organization (UNESCO)] as "Broad domain, branch or area of content covered by an educational programme, course or module. Often referred to as a 'subject' or 'discipline'." Retrieved from the website of UNESCO on 19.7.2017. In this context 'field of study' and 'discipline' are used interchangeably.

2 For further reading, see Mine Yazıcı' studies:
 a) (2011) Research in Translation Studies, from Interdisciplinarity to Transdisciplinarity and
 b) (2015) New Pathways in Research in Exploring New Concepts and Issues in Social Sciences Research.

3 The tendency towards literary translation may result from the supposition that the translators might practice their creative and linguistic skills more in the target language.

'literal' and 'non-literal' (Sager's classification in Olohan, 2011, p. 40). The latter is also classified, on a general basis, as 'technical' or 'scientific and technical' [In her doctorate dissertation regarding translation of specific fields, Elif Ertan (2010) remarks that 'scientific and technical translation' is also used as a subfield].

Along with this grouping, a further classification is enumerated in accordance with the branches of sciences including Social Sciences, Natural Sciences, and so on. In the narrower sense, a similar classification may also be made within the framework of specific disciplines which may be basically and duly identified in accordance with their 'branches of knowledge or teaching' and their field-related terminology or 'Third Code'[4].

The classification according to subject domains seems to be a preferred guidance in forming a frame for the training and/or research of specific field translation and interpreting. No matter how clear-cut such a classification is, a more comprehensive explanation or clarification is required for the right perception. For the sake of pursuing a standard approach, the classification of ISCED (International Standard Classification of Education) is taken into consideration herein because Yüksek Öğretim Kurulu [the Council of Higher Education in Turkey] states that this classification has been and is being used at both national and international levels[5].

From the perspective of Translation and Interpreting Studies, the difficulty of this discipline-based classification, on the one hand, may fall within the interdisciplinary relations/interactions. For instance, the field of 'Insurance' embodies not only the subject-matter and terminology of insurance but also those of other fields such as engineering, health, law (both national and international), and so on. On the other hand, using merely the generic name of the disciplines (such as Business Administration, Economics, Law, and so on) might also be perfunctory; for instance, the classification of activities/texts in terms of the discipline of Business Administration. This field which comprises such subfields as organization-wide accounting, finance, management, marketing, and so forth may sometimes be inadequately addressed as 'commercial translation'. This usage is likely to limit the scope of the field because 'commerce' might be assumed to focus merely on 'buying and selling goods and services'. Besides, as Olohan states (Ibid., p. 41), the term 'commercial' comprehends the idea of 'advertising' and/or 'translation

4 For further information concerning 'Third Code', see Mine Yazıcı's study presented in this collection.

5 This piece of information is given under the heading of 'National Qualifications Framework for Higher Education in Turkey'. Retrieved from the website of YÖK on 19.7.2017.

services rendered for payment' as well[6]. In addition, 'Business Administration' might sometimes be confused with 'Economics'. Nonetheless; both fields are listed separately in ISCED's classification of 'Narrow Fields of Education and Training'[7]. Likewise, employing 'Philosophy' as a part of Social Sciences may be accepted as another oversight according to ISCED's classification since it is considered within the field of 'Humanities'.

Briefly, ISCED's classification may primarily provide a standardization in categorizing the specific fields in TIS. This classification may easily be attained and while the narrow or detailed fields are used consistently, such confusions as using an application of a relevant field instead of the field name and the like could be prevented.

As academics, we should try to improve our programs in line with the global expectations and new orientations and it is of utmost importance to take the humanitarian, social, and sectorial needs into consideration. Therefore, the following suggestions may be made to meet the demand for specialization at different levels:

- practising communicative and pragmatic functions of specific field translation and interpreting in undergraduate programs,
- offering graduate and/or post-graduate translation and interpreting programs (involving courses on sign language interpreting as well) designed for the specific field specialists who are expected to deal with highly specialised texts/high contexts[8] (in the fields of health, international relations, and law, for example) where *well-formedness* should be assessed for the translation quality while it is vital to transfer the meaning accurately/precisely and free of error [which may cause devastating (even fatal) consequences otherwise].

Apart from the universities, such collaborative practices are already being encouraged in different sectors. For example, in the insurance sector, one major

6 The phrase 'Business Translation' may be preferred instead, because in ISCED's classification, one of the Fields of Education and Training is 'Social Sciences, *Business* and Law'. This economical and to some extent proper usage seems to cover all the subfields of this discipline. Nevertheless, it should be noted that the other Fields of Education and Training, namely 'Social Sciences and Law', are also used separately and accordingly. (For further information see, ISCED's classification on Detailed Fields of Education and Training regarding Business).

7 'Economics' (like 'Sociology' and 'Psychology') is one of the detailed fields of 'Social and Behavioural Sciences', whereas 'Business Administration' is listed under the detailed field of 'Business and Administration'.

8 as described by Hall (1976) in Byrne (2012).

reinsurance company based in the European Union employs experts specialized in the field of linguistics and terminology (Angı and Kerç, 2017).

This collection named 'Translating and Interpreting Specific Fields: Current Practices in Turkey' could be presented to our academic world thanks to the contribution of six distinguished academics, who have conducted the studies on the (sub)fields of engineering, international relations, medicine, sports in different modes as well as the linguistic approach to technical communication, which also provides an insight into the current practices on specific field translation and interpreting in Turkey. To raise more awareness in Turkish practices from the transdisciplinary perspective, it would have been a miss if 'Turkish Sign Language' were not included in this collection. This part has been translated into English by one of the Ph.D. candidates.

As a final note, it is hoped that this academic effort will serve as a compass leading to further studies on specific field translation and interpreting.

Wishing that this collection will serve to meet the social need for the improvement of our native country as well as our native language which deserves to be used more extensively with its rich and expressive vocabulary…

References

Angı Ayşegül and Kerç Elif. (2017). *Çeviriye İlişkin Eylem Kuramı ve Skopos Kuramı Çerçevesinde Özel Sigorta Poliçelerinin Çevirisi* [*Translation of Insurance Policies from the Perspectives of Translotarial's Theory and Skopos Theory*]. Çeviribilim ve Uygulamaları [Journal of Translation Studies]. Sayı 23 [Volume 23]. Ankara: Hacettepe University (published soon).

Ertan Elif. (2010). *Uzmanlık Alanları Çevirilerinin Tarihsel Süreci, Eğitimi, Önemi ve Çeviri Uygulamalarındaki Konumu* [*Translation of Specific Fields: History, Education, Significance and Fields of Application*]. Doctorate Dissertation. Marmara University. Institute of Educational Sciences.

Hall Edward T. (1976). Beyond Culture. New York: Anchor Books. In J. Byrne. (2012). *Translation Practices Explained, Scientific and Technical Translation Explained.* London and New York: Routledge Taylor & Francis Group.

Olohan Maeve. (2011). *Commercial Translation.* In Mona Baker and Gabriela Saldanha (eds.) Routledge Encyclopedia of Translation Studies. 2nd Ed. London and New York: Routledge Taylor & Francis Group.

Yazıcı Mine. (2011). Çeviribilimde Araştırma, Disiplinlerarasılıktan Disiplinlerötesine [Research in Translation Studies, from Interdisciplinarity to Transdisciplinarity]. Istanbul: Multilingual.

Yazıcı Mine. (2015). *New Pathways in Research in Exploring New Concepts and Issues in Social Sciences Research, Chinese and Turkish Perspectives*. New York: Untested Ideas Research Center.

Electronic Sources

UNESCO

http://uis.unesco.org/en/glossary-term/field-education

YÖK's 'National Qualifications Framework for Higher Education in Turkey'

http://www.yok.gov.tr/en/web/uluslararasi-iliskiler/isced

Prof. Alev Bulut Ph.D.

Istanbul University
Department of Translation Studies

Foreword

Translation and Interpreting Studies (TIS) as a scientific discipline undertakes the tasks of describing, analysing and categorizing the practical data of translation and interpreting of any content -fictional or non-fictional and technical as two legendary poles of the past- in search of a systematic study of theoretical frames and universals. Technical content, unlike the long-scrutinized and studied literary contents and text-types, still poses us, the scholars, with many unanswered questions as the most challenging content category in TIS due to its requirements and impositions of background knowledge in domains such as Medicine, Law, Economics, and so forth.

Although the most common and widespread forms of translation and interpreting practices fall into the main categories of Physical and Social Sciences, basing on the heavy load of the domain translation practice carried out at translation bureaus (official and technical documents of operational nature for the most part) as well as the hectic traffic of interpreting tasks in many domains performed at conferences and meetings, domain translation cannot be said to be a pioneering sub-field of translation research or translator training. Mary Snell-Hornby (1988/ 1995) is a pioneer in our field in terms of providing us with an integrated approach to translation theory and practice instead of partial ones which led to the description of the sub-field of translation with technical content under the integrated umbrella of translation research. Although interpreting, as an aspect of spoken language, was not then associated with technical content, interpreting studies today is integrated with the general trends and turns in the field.

The theses and publications on literary translation initiated the theoretical introductions to the scientific field, 'Translation Studies', in the academic literature in the world and in Turkey. The widespread practices in the technical/ scientific/ Language for Specific Purposes (LSP) based translation, however, took time to be scrutinized as part of the intellectual background and corpora of our field.

The factors underlying the above observations are many-folded, indeed. As a scholar and practitioner based in Turkey, I will try to summarize the outlook of domain translation and interpreting in theory (research) and application (training and practice) from my local perspective into four points, thinking that it might apply to other observations to be made elsewhere for other regions or countries:

- Firstly, the scientific and/or technical content requires background knowledge in respective fields of expertise. The practice in most contexts of translation and interpreting (putting aside the fictional and daily communication based-contents) takes technical background knowledge for granted since the majority of the material and subject matter of the profession are technical and/or scientific in content. The knowledge and terminology of the domains, thus, become critical and indispensable aspects of specialized translation and interpreting, which poses a major challenge in the training of translators and interpreters.
- Secondly, the universals or norms of practice in different domains (medicine, law, economics, and so on) always seem to be out of the scope of interest or reach of the academicians of our field. In academic translator training, it has always been difficult to recruit experts from the field to conduct the courses with technical content, let alone the difficulty of making the curriculum of these courses sustainable enough in trainer and curriculum wise. This created a gap between the practice and practical training of the fields in question, and the theoretical studies and researches conducted on them. The experts of the field were not academicians of Translation and Interpreting Studies and the academicians of the fields of translation and interpreting have not always been practitioners themselves, providing us with hands-on research of their practices.
- Thirdly, domain translations require interdisciplinary perspectives not only in the study of the content and concepts of the domains but also in the study of the complex text-types and registers of the technical fields. Technical writing and transfer skills require communicative skills that are described as the purpose of the task in translation and interpreting. The more interdisciplinary our perspective gets, the more insights into the typologies of texts (registers, styles, and discourses) we will adopt to notice the wide range of text-types between the two poles of use (from the most official/ scientific/ technical to the least official/ technical/ popular/ daily) in major domains (Medicine, Law, Economics, and the like) and fields of Engineering based on Primary Physical (Physics, Chemistry, Biology) and Social (Logic, Sociology, and so forth) Sciences.
- Finally, having listed the above points of issues and challenges, we are left with many text-types between two poles of style from scientific articles, technical reports, official/ legal documents to text-books, manuals, popular science texts and many other hybrid types that have found their own functional places in the everyday uses and operations as the need occurred. Accordingly, different

target reader and user profiles and purposes of translation created different facets for the text-typology of the texts produced in the same domain.

We can say that the integrated and generalistic approaches in TIS make it easier today to include the Technical Fields/domains to the theoretical discussions of the field since (unlike partial approaches of the past) today's translation theories see the field as a whole; thus, they can be evaluated from any perspective the researcher chooses to employ, from the linguistic perspectives to the cultural and functional trends of TIS. We have really gone a long way since James S. Holmes's (1975) seminal paper 'The Name and Nature of Translation Studies' [in L. Venuti (ed.) 2000. pp. 172–186][1]. Now that the foundations are laid, we are ready to enlarge and update our horizons in the scientific analyses of the translation and interpreting of technical content.

The current collection titled **Translating and Interpreting Specific Fields: Current Practices in Turkey** is a pioneering study, most probably the first one of its kind in our case bringing together the major aspects of the Technical Fields (namely, text-types/ discourses/ terminologies and technical aspects) and descriptions of translation and interpreting practices (including sign language) from the domains of Medicine/ healthcare and Diplomacy to Sports as a contribution to the worldwide family of TIS academia.

If specialized language produces technical content, the transfer of technical content will naturally require specialization and expertise. I would like to add a few more observations on the topics of the works in the collection to briefly salute each of the articles with regard to their focus of specialization.

* Medicine is a major domain of professional practice and training. Medical translation can be seen as an extension of technical and scientific specialization in the field of medicine. It is also considered as a sub-division of community interpreting under the title of healthcare interpreting based on the emerging contexts due to the international developments (such as the recent Syrian refugee crisis in the Middle East). Another professional context for prospective graduates of translation and interpreting divisions of universities, in this sense, is the health tourism serving largely for medical tourists as an emerging sector in our country.

* Technical texts are basically the products of the informative function/use of language which does not mean to discard the huge category of operative texts such as manuals whose function is to tell or show us, consumers/customers, how

1 Also see M. S. Hornby's (1997) *The Turns of Translation Studies* for a chronological discussion of the respective turns in TS.

that information can be of use. Translation of manuals is an industry in domain translation as the most visible, useful, and public aspect of the field. Technical translators are writers of technical discourse producing texts just as the producers of the source texts of those translations do, which means they are experts, too.

* The complex and many-folded practice of the translation of specialized [or LSP as referred to by a line of research, for example Margaret Rogers' *Specialized Translation Shedding the 'Non-Literary' Tag* (2015)] content requires us, the translation scholars, to think at an interdisciplinary dimension where Translation Studies and Linguistics/Discourse Analysis stay in contact. The study of the strategies of specialized text production in any language (as source or target text language) offers valuable linguistic and textual hints for the transfer of that specialized content with a purpose.

* Diplomatic content is not only challenging for translators and interpreters with its requirement of skills but also quite challenging for TIS scholars to conduct research on its written and spoken settings of practice. It is as technical as any other informative context of transfer with its requirement of competences in the fields of Social Sciences (or in the narrower sense beginning with Politics, Philosophy, and Sociology). Moreover, diplomacy poses challenges in terms of implicitness and cooperation as do the other addressive contexts of communication making use of daily and cultural background, which is added to the technical and specific/domain-based difficulties of the practice.

* Interpreting is conducted between two languages, and in the case of sign language the practice is carried out between a verbal and a signed language. The whole accumulation of knowledge and theoretical background employed for the explanation of any practice and product of translation and interpreting is definitely and entirely valid for sign language interpreting. Sign language interpreting is mostly technical in content – just as any other interpreting practice is – which is easy to observe from the contexts and contents of meetings and conferences where sign language interpreting is required.

* The wide range of contexts and settings of interpreting is a vast field waiting for the interpreting scholars to explore, describe, and categorize under the theoretical frames of interpreting research. In our local field, one of the emerging contexts in the consecutive mode of interpreting is Sports (mostly the football/ soccer). The description of the settings and tasks of sports interpreting is an emerging topic in TIS research attracting the attention of those young academicians (and not so young ones like myself as well!) who are interested in several fields of Sports as

players, referees, or fans. Sports discourse and interpreting in the field of Sports as technical content is as challenging as any other interpreting practice.

Looking at the number of academic publications and field research produced in domain translation, it can be stated that we are at a cross-roads in the study and teaching of translation and interpreting of technical content in its own right under the integrated umbrella of TIS approaches and methods that have proven their validity over the years with written / oral / multi-modal / audio-visual / literary / non-literary/ technical data.

The articles of the collection, written by TIS scholars and researchers in Turkey, reach you in English/ the lingua franca of academia displaying the authors' and the editor's intention of introducing the recent developments in the technical/ domain-based practice in the field of translation in our country to the rest of the world. The editor of the collection Assist. Prof. Ayşegül Angı Ph.D., as a domain specialist herself, deserves a big round of applause for setting out to bring together the hand-picked articles on a topic and field of research that is still in its infancy in our country.

It also touches me deeply to be writing this foreword for a publication of Marmara University where our beloved colleague and friend Elif Ertan Ph.D., a very promising young scholar whom we lost very tragically and unexpectedly in 2014, got her doctoral degree with her thesis (published in 2013) on the historical process, teaching, importance, and applications of domain translation in Turkey.

I hope and sincerely believe that the collection in your hands will trigger the publication of many more collections on the theory, practice and training of domain translation making the local practices meet under the umbrella of major theoretical approaches in TIS.

References

Ertan Elif. (2013). *Uzmanlık Alanları Çevirilerinin Tarihsel Süreci, Eğitimi, Önemi ve Çeviri Uygulamalarındaki Konumu* [*Translation of Specific Fields: History, Education, Significance and Fields of Application*]. (published PhD thesis). İstanbul: OFIS 2005 Yayınevi (Publishing House).

Holmes James S. (1975). The Name and Nature of Translation Studies. in L. Venuti (ed.) (2000). *The Translation Studies Reader*. London: Routledge.

Rogers Margaret. (2015). *Specialized Translation Shedding the 'Non-Literary' Tag*. UK: Palgrave Macmillan.

Snell-Hornby Mary. (1988/ 1995). *Translation Studies: An Integrated Approach*. Amsterdam: John Benjamins.

Snell-Hornby Mary. (1997). *The Turns of Translation Studies*. Amsterdam: John Benjamins.

Prof. Aymil Doğan Ph.D.

Hacettepe University
Department of Translation and Interpreting

Interpreting in Diplomatic Contexts

"big victories are measured in small gains of verbal territory"

Abstract: The study dwells mainly on interpreting in the diplomatic environment, which is a crucial aspect of international relations since these relations are the steering wheel of interactions in the fields like politics, economy, ecology, which are the world issues of vital importance, having direct impact on world peace and wars. The term 'international relations' bears the ideas of reciprocity, communication, connectedness among the negotiating parties. Policy is a determining factor in the decision-making process of international relations, diplomacy is the tool of interaction that enables these decisions to be implemented through peaceful means and protocol rules maintain the smooth environment for this interaction. Diplomacy, which dates back to ancient times, can be described as the art and science of managing international relations through recognition of the intention and the hidden baggage, making use of intuition and the merits of diplomatic jargon and discourse. Implicitness in discourse provides not only politeness but also the conditions for clever manoeuvring, which is also necessary for behaving in the way that fits human dignity. Universe of diplomatic discourse, therefore, covers a wide range of areas, some of which encompass the knowledge and understanding of international relations in both general and specific sense, the events in historical development and persons that have an effect on the situation in question; the discourse to be employed in these areas; clever use of language in at least two languages, mother tongue and a foreign one both in the semantic and pragmatic sense; the polices of the parties making interventions; background of the policies; the relevant legislation; the intention and future plans of the parties intervening; the protocol rules and codes of conduct. Diplomacy and interpreting in the historical sense emerge and develop almost together as both are indispensable aspects of intercommunal and international relations. In addition, just as the diplomats who were the persons competent in policy and diplomatic relations were also good at languages, thus authorized to establish the liaison, the interpreters who were the persons competent in languages were authorized to carry out international relations and thus, diplomatic negotiations. It is noteworthy that both were privileged professions during the Ottoman Empire, and the interpreters of different origins, were also high rank officials; for example, one being appointed to collect taxes or another signing a treaty in the name of the Empire. Education of diplomatic interpreting should include not only some part of the field knowledge for the diplomats but also discourse analysis with specific reference to cooperative principle, notion of truthfulness condition, seven standards of textuality, register theory, conversational implicatures, relevance theory, as well as source and target language oriented studies of translation and interpreting. Implicitness and politeness are of crucial importance to keep the diplomatic wheels turn smoothly.

1. Introductory Remarks

International relations have been of crucial importance in establishing the past and the present of our world and will continue to be so in the future as it maintains the perpetuation of the world's existence. The term 'international relations' presupposes the idea that there is 'reciprocity' in the act of establishing relations at the international level. Reciprocity is among one of the 'felicity conditions', as described by Hurford and Heasley (1987, p. 251), that must be fulfilled in the situation in which the act is carried out, if the act is to be carried out properly or felicitously. For example, felicity conditions for the 'act of promising' are as follows: "The speaker must have the intention to carry out the thing promised and the thing promised must be something that the hearer wants to happen"; or for the 'act of apologizing' are as follows: "The speaker must be responsible for the thing apologized for and the hearer must not want the thing apologized for to happen or to have happened." In the same vein, the 'act of establishing a relationship' can only be fulfilled if the sides come together with the 'intention' to be in contact and get involved in an exchange of subsequent reciprocal acts. In the diplomatic context, this exchange must be carried out on equal footing. The interlocutors of the meeting must be counterparts; especially the host interlocutor must not be at a lower rank than the guest interlocutor, yet may be at a higher rank. The officials of foreign relations, like diplomatic missions, consular posts, are the representatives of their countries, thus, must act according to the diplomatic codes of conduct so as not to cause any crisis. Once, the crisis between Israel and Turkey came into being on 11.1.2010 because the Ambassador of Turkey was seated on a lower chair in the Office of the Deputy Minister of Foreign Affairs[1]. However, the initiator of the crisis is the intention rather than the act itself.

Intentionality is virtually the primordial factor that gives direction to almost all acts. Searl (1983) in his book titled 'Intentionality' puts forth the relationship between the intention in the mind and the expression of this intention through speech acts and actions. Thus, the realization of intention bears some stages. Firstly, there develops the 'prior intention', which bears the propositional content in the verbal mind; secondly, prior intention leads to the 'intention in action' or as Searl also puts it in terms of a different perspective, 'intentional action'.

1 Retrieved from the website of Haaretz on 12.3.2017.

Intentional action is the bodily movement manifesting what is in the mind to be perceived visually. Thus, in diplomatic context intentionality can be regarded as the backbone of relations.

Intentionality is also listed within the scope of textlinguistics by Beaugrande and Dressler (1981), among the 'seven standards of textuality', referring to the formation of a text with a communicative value. These standards set, firstly, function to describe the characteristics of a well-formed text; secondly, any sequence of sentences to be regarded as 'a text' must fulfil some conditions, namely 'intentionality', 'cohesion', 'coherence', 'informativity', 'situationality', 'intertextuality', and 'acceptability'[2]. In order to elucidate what these standards are for, the verbal text produced in the target language by the interpreter can be taken as an example. The verbal text produced must be acceptable by the target hearer(s). This condition is only fulfilled if the language of the interpreted verbal content is well-formed and fluent, which also depends on the proper realization of all the other standards. *Well-formedness* includes aspects such as the use of relevant jargon, relevant register, relevant style, relevant syntactic sequence to convey both what is said and what is not said or concealed or blurred. *Cohesion* refers to the structural well-formedness; however, meaning assembly in the verbal mind can occur by means of the connectors. There are cases when the speaker does not use any connectors between the two utterances but expresses the idea through body language, which cancels the effect of the former. In this case, the interpreter must decide as to which connector is the proper one to convey the right message, making the text into a logical whole. This means that 'coherence', which establishes the logic and connectedness between the syntactic sequences, is as important as 'cohesion', to build up a text with communicative value. This connectedness is also maintained when the speaker relies on other texts to lay the grounds or to remind the content thereof during the delivery of the speech, which refers to the standard called 'intertextuality'. The reference to other texts may be subtle, yet the traces of them precede or follow the text being produced. Another standard, 'informativity' is also very important in the formation of a text. During the question and answer session, extemporaneous interventions from the audience may contain utterances which are irrelevant and redundant. Redundancy is one of the challenges of the interpreter which requires utmost attention when making decisions about whether to convey all what is said or not. Redundant phrases may confuse the speaker's mind at the rostrum who is listening attentively, focusing

2 For further information, see Doğan (2014, pp. 112–117); Beaugrande and Dressler (1981).

on the point raised so as to provide the best reply; thus, it is the task of the interpreter to put the loose discourse into a coherent and comprehensible one. However, not all international meetings are appropriate for this. There are cases when even the hesitation of the responder is evaluated as a message especially in health and diplomacy. Thus, in oral communication what is 'informative' differs from situation to situation. 'Situationality' is another standard which maintains that the verbal text is relevant to the time and place it is produced in. All of these six standards function harmoniously only when the 'intentionality' factor, the seventh, is not disregarded. The interpreter must be aware of the fact that the only means to provide communication between the parties with language barriers is the interpreting rendered; thus, the interpreter should act with the intention to convey the reproduced verbal text to the speaker at the rostrum in such a way to enable comprehensibility. During the course of interpreting, *intention* is made clear through prosody, stress, intonation mostly but also all the other standards affect one another and the whole.

The act of 'communicating' also entails 'association' and 'connectedness' creating action as opposed to its antonyms such as 'dissociation' and 'disconnectedness' giving rise to the inertia of the act of stillness, thus non-action. The collocated prefixes such as 'com-', 'col-' and so on, and the words such as 'interaction', 'communication', 'collaboration', 'international', all semantically entail the positive attitude of moving towards one another and to act together with one another, to prefer peaceful negotiations even when there is tension between the states. In short, the relations between the countries start with their intention and good will to establish a communication environment based on the policy of each making use of diplomatic means.

2. Policy and Diplomacy

International relations are established on two foundation stones: *policy* of a country – be it internal or external as each relies on one another – and *diplomacy*, which is virtually international.

Policy can be described as adopting an attitude as to the management of the internal and external affairs of the country regarding the underlying intention of the government in accord with that of the Grand National Assembly as well as realizing the goals following the relevant road maps; taking into consideration de facto and de jure conditions; making decisions and putting these into practice; and maintaining their applicability and sustainability.

Diplomacy, on the other hand, can be described as realizing the external political goals at the international level but with peaceful means, making use of relevant strategies, methods, and discourse. It goes without saying that each country wants to improve its position and achieve its goals that will provide betterment in the present situation and to open horizons for the future using a variety of means; however, this is maintained only with diplomatic means, namely, through creating a peaceful ambiance and interactive environment, and using the sort of discourse which can be referred to as 'diplomatic politeness'. Thus, diplomacy has a special style of language use and a way of addressing sensitive problems. As mentioned above, the intention to maintain and preserve the peaceful negotiation atmosphere is of critical importance, imposing crucial responsibilities on the negotiating parties. The interlocutors of the negotiation, in other words the parties interacting, are expected to use the sort of discourse acceptable by the international community, which reminds us of another standard of textuality, *acceptability*. For example, a ministry of a developed country may get into contact with that of a developing country to provide better living conditions. The style and wording of a text to make an offer will be different when the same content appears in a letter, in an agreement, or in a business contract. Similarly, the discourse of the oral communication during the first encounter will be different from the one preferred during the meeting to negotiate the terms and conditions. Thus, the intention, the goal, and the content may be the same but the text type is the most important determinant of an acceptable preference of discourse. A translator or an interpreter, on the other hand, is likely to come across with such a situation when the source text, be it written or oral, has not been drafted with an appropriate wording and style; then he/she may take the initiative, informing the text-writer to restate the content in the required wording and the style. In oral communication, the interaction is a live one; thus, the words of an utterance have a different impact than the ones in a written text when read silently. An utterance is a different one each time it is expressed; whereas a sentence is the same each time it is read silently because with each utterance, even though the wording is the same, the intonation, the stress, the pitch change as the intention of saying starts to change from an utterance to the next utterance. In oral communication, the function of an utterance is perhaps more apparent as the intonation makes its contribution to the sense and the realization of the sense in vivo[3]. For this reason, Jönsson and Hall (2002, p. 1) underscore the functionality of words, stating that "saying is doing" and "doing is saying". This idea was theorized by Austin's Speech Act Theory, which, in brief terms, draws the attention to the function of what is said or written. For example, the utterance by an Ambassador "Our allies think

3 For further information, see Hurford and Heasley (1987, pp. 15–30).

that your country shares the same values with this organization" may mean exactly what we understand considering the denotative meanings of the words comprising it; in this case, the function of this utterance is 'making an assertion' or 'giving information'. However, it may mean something completely different from what it seems to say; for example, it may be a criticism saying "…but you do not behave in the expected way"; or a threat saying "…if you do not conform to the rules, we will be obliged to take an action" or a memorandum mandating "..it is not good for you to work with that organization, you should stop it"; or a help offer saying "… if you have any difficulties in keeping up with the values stated, we can discuss and help you to comply your legislation with ours in line with these values" or else. Thus, especially in diplomacy the wording and the style should be arranged much more carefully and meticulously, considering the functions of utterances and sentences to keep the diplomatic wheels turning smoothly[4].

Etymologically 'diplomacy' comes from the Latin adjective *diplomaticos*, which refers to 'diploma (genitive diplomatis): official document conferring a privilege'[5]. Hence, bearing in itself the semantic entity that diplomacy is an official task, the act of diplomacy requires special training to be crowned by a diploma; thus, the doer of this act is a privileged one. A modern description as to where and how it takes place is made by Barston (1997, p. 1) who states that diplomacy is carried out through one to one communication in private settings or in the form of exchange of views in various mediums, also making use of lobbying and paying visits; and the discourse can include recommendation or threatening. Tezok (2014, pp. 4–5) describes 'diplomacy' as the art of establishing communication. In addition, Matteucci (2001, p. 64) posits that diplomacy is 'an act of recognition' first; in other words, 'an acknowledgement of realities that surround the process, in particular the hidden baggage,' and is also 'an act of intuition' that will help 'find the way in a stark landscape', where there is no certainty of success in the end since the diplomat is vis-à-vis many unexpected adversities that pose challenges, yielding the undesired results and turning all efforts vain, which makes one think that it is perhaps this act of venturing in addition to the other factors that attracts people to be diplomats. From this point of view, diplomacy and interpreting resemble one another. Risk-management strategies are likely to include anticipation. In the

4 Two newly published books can provide insight into the world of diplomacy through first hand anecdotes: 'Diplomatic Discourse', presenting the interviews by the US Ambassadors from around the world during summer 2013 edited by Schuster and Stern (2015) and 'Bir de Bizden Dinleyin [This Time Listen from Us]', presenting the memoirs of Turkish diplomatic wives by Onat (2016).

5 Retrieved from the website of etymonline on 2.8.2016.

same vein, it can be postulated that anticipation in both interpreting and diplomacy includes not only prediction but also intuition. The road taken during the rendering of interpreting is a rough one, paved with obstacles and challenges to be overcome in very small-time periods like seconds, where success in one moment is likely to end up with failure in another moment due to the challenging factors such as immediacy, simultaneity, and incrementality, as described by the general public and formulated with terms by Setton (1998, p. 165). Due to the fact that interpreting is a mental endeavour and skill developed within the mind, integrated with various features of personality, it is hard for the ones who have not experienced it to well appreciate the challenges. However, risk-takers learn how to deal with it and have a knack of it in time. As Doğan (2015, p. ix) stated in a previous publishing, interpreting is an adventure to be experienced, while longing for the precious moments to be savoured during the process.

From a different perspective, diplomacy is a crucial act through which decisions are made to have effects on the lives of a large number of populations. Top level diplomats are the *ambassadors*, who represent the president of one country in another. An *embassy* is the representative office of the foreign country, established in the capital city of a country. The office in cities other than the capital city is called *consulate* and the appointed high-ranking officials in consulates are called *consuls*. These diplomatic posts have privileges and immunities in the host country. The agreements[6] ensuring these are the Vienna Convention on Diplomatic Relations of 1961 and the Vienna Convention on Consular Relations and Optional Protocols of 1963, and Guide to Diplomatic Missions in Turkey. While the *politicians* are responsible for the international relations in the political sense, the *ambassadors* and *consuls* are responsible for protecting the rights of their own citizens in the host country and for furthering the economic and commercial relations between the states.

Behaving in the diplomatic way is not a style pertaining only to diplomats; diplomatic approach is needed in many other professions as long as peaceful and polite negotiation is concerned and thus provides a broad area to study in the academic sense. The concept of *diplomacy* is more comprehensively described herein by Doğan as "*the art and science of managing international relations through successful conduct of negotiations carried out in peaceful means and polite terms making use of the merits of diplomatic jargon and discourse, enabling clever manoeuvring to maintain, to secure, and to protect national benefits in conformity with*

---

6 The *agreements* retrieved from the official website of the Ministry of Foreign Affairs of the Republic of Turkey on 2.1.2017.

the wording and spirit of the constitution of the country as regards national dignity, in an environment arranged by the protocol rules."

Diplomacy is the tactful management of international relations and it is prone to disputes stemming from the sociocultural differences. The main task of diplomacy is to minimize the sociolinguistic differences; to come to terms to solve the socio-political problems. Cremona and Mallia (2001, p. 305) draw attention to the fact that dealing with international relations requires specially trained professionals, namely the diplomats, who skilfully use the diplomatic language, and when translation is needed, translators and interpreters are in action. The most rewarding part of being a translator or an interpreter in the field of diplomacy is perhaps that they have significant contribution to the formation of history.

3. Diplomacy and Translation/Interpreting: A Brief History

The history of settlement in Anatolia was found to date back to 9600 BC when the pre-historic ruins were discovered in Göbekli Tepe. Translation and interpreting activities in Anatolia seem to have started when the communities speaking different languages got into contact for some reasons like trading, which dates back to the times of civilizations such as Sumerians, Assyrians, Babylonians, Hittites, Greeks, Romans, Seljuks, and Ottomans. Abu Jaber (2001, p. 52) states that the most ancient diplomatic language used was for a peace agreement, which was found to have been inscribed on a clay tablette in 2400 BC and is displayed in Damascus Archaeological Museum. The first diplomatic translation was the famous Kadesh Peace Agreement, which was inscribed in 1270 BC (one being in Hittite language and the other in Egyptian language), and the masterpieces are exhibited in the Istanbul Archaeology Museums. Since the contacts between the communities of different languages and interpreting activities started as early as those times in Anatolia, it is called 'the cradle of civilizations'.

From the etymological point of view, the word 'interpreter' is said to have been derived from the word 'latimer' in English and 'latimier' in French, referring to the person who speaks Latin, the 'speaker of Latin'; yet the assigned meaning was the 'corruption of the Latin language, which was originally graphic'. Moreover, in Sanskrit '*prath-*' means 'to spread abroad' and 'interpreter' from Old French is 'to expound the meaning of, render clear or explicit'[7]. In addition, Akkadians referred to this act of mediating between the languages as 'targumannu'. Interestingly, it was also stated that this word was originally derived from the Semitic

7 The definitions were retrieved from the website of etymonline on 2.8.2016.

root 't-r-g-m', which means 'word'. The 't' at the beginning refers to the causative form; and 't-r-g-m' means 'causing to be a word' or 'turning to be a word'. Aramaic translations of the Bible are called 'targums' from the same root (Hoffman, 2004, p. 167). In various languages, the word was transliterated into various forms in various languages, for example, 'tarjumān', 'targemana', 'tercüman'. In Hebrew 'tirgem' is the verb form, and the noun form 'targum' is 'translation'; whereas 'meturgeman', 'turgeman', 'ha/metargem', and 'targeman' refer to the translator. In Latin, the meaning of 'interpretator' and 'interpretatio' is 'to explain a word or statement with another word or statement both in the same language and from another language'. In Hebrew, the difference between spoken and written mediation was given with the variation between 'metargem' [the translator of written texts] and 'meturgeman' [the interpreter of speeches] (Alexander, 2014, pp. 229–230). In Latin 'interpres' is the *expounder*, the person who explains what is obscure (Pöchhacker, 2004, pp. 9–10).

Turkish records demonstrated that official appointment of two dragomans and two translator clerks took place as early as the reign of Alaaddin Keykubat I (1220–1237). It was in 1479 when the 'imperial dragoman', Lutfi Bey, was designated to sign a peace treaty in Venice. The first group of people assigned as the mediators between two speech communities were the tradesmen who learned foreign languages during their journeys and in time the number of the tradesmen who could speak foreign languages increased. Towards the end of the 15th century in Europe the interpreters who were specialized in diplomacy started to mediate in commercial and diplomatic transactions in the large area called 'Levant' which extended from the east of Italy as far as the whole eastern Mediterranean and the Middle East. Thus, these interpreters were also specialized in the Ottoman and Islamic cultures as well as the languages used in this area. The famous sailor Jacques Cartiér took the first step to solve this language problem and took two native children away from their tribes to France to train them; however, separating them from their families for his own benefit was not considered humane. Later the training was provided in a more systemized way, for example, the founder of Quebec, Samuel Champlain established an institute in 1608 and provided training, by indwelling the adventurous, young (between 8–30 years old) French people into the native tribes which they traded the most so as to enable them to learn the language and culture in a natural environment as the natives did. In the 16th century, this method became widespread through the countries and diplomatic interpreters came into being first in Europe, specifically in the Venetian Republic. In this way, not only were they experiencing the Renaissance but also, with their fleet, they were trading with the countries in the Mediterranean.

Having accommodated in a foreign country, generally oriental in the neighbourhood of the Ottoman Empire, to master the language and culture of that country, these young people became mediators of communication. They were called 'giovani di lingua' by the Venetians. This name was translated into 'jeune de langue' and 'enfant de langue' in French, 'joven de lenguas' in Spanish, 'sprachknabe' in German and finally 'dil oğlanları (language boys)' in the Ottoman Turkish. These people later started working as professional translators and interpreters at the administrative offices, embassies, and consulates (Cáceres-Würsig, 2012, pp. 127–144; Doğan, 2012, pp. 57–72; Doğan, 2015, pp. 10–12; Hitzel 1995; Paker, 2009, pp. 550–559; also see Diriker, 2015, pp. 89–107 and Timur-Agıldere, 2010).

The archives in the Ottoman Palaces demonstrate that the Ottoman Empire was different from the other states, most of which were ruled by monarchy, especially in that the Sultans of the Empire never intervened with the religion, language, and culture of the states which were under the rule of Ottomans, or never attempted to do what the imperialist countries did during the colonial period. What the Empire did was to merely collect taxes from the states within its territory but never exerted any sort of force or power to change the religion, language, or culture of the peoples under their reign. This is why the peoples in three continents had the chance to preserve their own identity, their own religion, and their own culture and language. The Ottoman Empire was composed of multicultural and multilingual people with different religions. For example, as Cáceres-Würsig (2012, p. 130) stated there were Greeks who were Orthodox Christians; Armenians; Sephardic Jews; Latins who were Levantine Catholics and were originally Venetian; Genovese; and Cypriot merchants. Muslims were in the position of authority, yet many people from the other religions and communities were also authorized at the top-level positions. Since Istanbul was a prominent trading centre of those times, the diplomatic network was needed to provide interpreting and translation services and to mediate during the communication process between the immigrants, merchants, and foreign authorities and the Ottoman authorities. This service was provided by the immigrant non-Muslims. Turkish was the official language; however, Arabic, Armenian, Greek, Italian, Farsi, and Slavic languages were also used extensively. In the 17th and 18th centuries the countries in Europe which were ruled by monarchy such as Austria, Russia, Prussia, Spain, and Denmark sent their envoys to Istanbul to open their diplomatic and commercial missions. Furthermore, the Sultan of the Empire granted the dragomans, who were the people belonging to minorities in the Empire, a document called 'berat' (a special certificate), which provided them with many additional rights, privileges, and facilities. The position of these dragomans mediating between both the Europeans and Ottomans was thought to be dubious in time as they had to protect the rights of both sides. Moreover, Venetians wanted to secure the dragoman's loyalty. Taking these as their starting point, Europeans decided to take more control of the education and they themselves chose the young ones to be sent to the target country /city (Istanbul, in this case) so as to enable them to learn the target culture and language, and to experience and master the diplomatic environment there. This system was first utilized by Venetians in 1551; they sent two 20-year-old notaries to Istanbul Embassy of the United Kingdom, who were the first 'giovani di lingua'; France followed Venetia and in 1669 education was launched in schools to train 'jeuenes de langue'. The first students were two Armenians but in the following years the French Government decided to educate the students of French origin as well as the grandchildren of the Levantins who had once settled in the East Mediterranean. Then the school, opened in 1795, became an obligatory education institution for the ones who were to be trained as 'jeuenes de langue'. In 1754 an academy was also opened in Wien and the students were sent to Istanbul to master their diplomatic skills after their apprenticeship training. These European young men got married to local girls and brought up new generations that were the synthesis of the cultures of both the East and the West. The Spanish merchants demanding the presence of an interpreter in their courts were granted the right to have an interpreter in the court, and thus, the interpreters in time became an indispensable part of the court and diplomatic negotiations. They were working under the body of the consulates or embassies, were permitted to sit at the same table to eat and could dwell in the same residence with the staff of the consulate or the embassy; they were paid very well and their travel expenses were covered. It was the envoy's task to provide information to the Sultan about the professional improvement of the dragoman. In 1785 the attaché, the expert of special fields such as commercial, military, and diplomatic affairs, took over this task when such offices were opened under the body of the embassy, which was later followed by other embassies. The opening of the attachés' offices necessitated the existence of more translators and interpreters for more specialized fields. This paved the way for a much better perception and prestige of the profession, enabling much better working conditions and finally the assigned tasks encompassed positions at the state level which went beyond just translating and interpreting (Cáceres-Würsig, 2012, p. 129–138). Therefore, dragomans were very important people and "most dragoman families made an effort to perpetuate the family profession as long as they could." (Diriker, 2015, p. 91).

Interpreters in the diplomatic medium were considered to be privileged and powerful in the eyes of the public as they were allowed to speak after the King.

The ones educated in Europe so as to establish communication with the Ottoman Empire were later authorized as ambassadors (Thiéry, 2015, p. 107). Within the course of the professionalization and institutionalization, efforts gained momentum together with the incremental need in the world which underwent many changes in the political, technological, and socio-economic sense; therefore, translation and interpreting services became indispensable for all the international meetings.

4. Spread of Diplomatic Relations and Developments in Interpreting

In an academic conference held in India and in its final publication titled 'Nation-building Through Translation', Doğan (2012, pp. 56–72) drew the attention to the fact that the translation and interpreting activities that took place in Anatolia and within the territories of the Turkish land all through the ages not only facilitated the communication between the interacting parties but also shaped the national consciousness by flourishing the culture and language of the successive states and helped develop a common mind among the peoples. Another underscored point regarding the interpreters and translators was twofold:

a) They were respected people with high prestige, therefore, granted many privileges.
b) Being the native speakers of foreign languages, mostly the non-Muslim people were educated to be appointed as interpreters or translators, yet later they gained higher status to the extent that there were such positions they occupied as the public administrator of the province, the tax-collector of the Empire, the plenipotentiary to sign a very important treaty in the name of the Empire, and so forth.

The positions they were allowed to occupy and the tasks to perform were the proof of the indiscriminatory, positive, and good-intentioned attitude of Turks towards the foreigners. Baigorri-Jalón (2014) wrote about the later stages of the professionalization process. The Industrial Revolution led to the geographical discoveries, which caused changes in the life styles. The wars during the rule of Napolion caused the movement of a mass of people. Vienna Congress, which was held in March 1814, was the largest international meeting ever held up till that time, where many countries convened to make socio-politically very critical decisions, which was crowned by the Treaty of Paris signed a month later to end the war between France and the Sixth Coalition. In this conference, the need for translation was much more than the one for interpreting. Among the other pioneering ones, this conference was a very significant one to be remembered, marking the history of translation and interpreting as a crucial milestone since it enabled the convention of a multinational meeting. In 1919 Paris Peace Conference, which was signed in the aftermath of the World War I, was another milestone marking the importance of particularly interpreting. In this meeting, French was the language of diplomacy; however, Wilson, the President of the USA and Lloyd George, the Prime Minister of the UK were not in good command of French language; thus, the interpreting service was provided by interpreters, one of whom was Paul Mantoux, a well-known interpreter. It was after the World War I that the USA started to bear an effective role in the international politics, which made it possible for English language to be accepted as an international language. Furthermore, English language was recognized by the League of Nations as the language to be used in conferences along with French language.

During the period between the two world wars, Stalin, Mussolini, and Hitler were using their unlimited powers, playing their roles in the political fora where one to one meetings were held between the other leaders. During these talks, the quality of the service of interpreting provided by the private interpreters was of utmost importance, and Paul-Otto Schmidt, Hitler's interpreter, was one of the good examples of the sort of diplomatic interpreting where confidentiality and trustworthiness were vital. In the aftermath of the World War II two crucial political events took place: one being the Nurenberg Trials and the other was the establishment of the United Nations. The Nurenberg Trials became famous also for the first use of simultaneous interpreting equipment. It should be noted that there was another device, hush-a-phone, used before this system but it was not specifically designed as an equipment for interpreting. The interpreter was located inside the bottom part of the rostrum where a window was opened and rendered the interpreting in a low voice through this hush-a-phone, a voice silencer originally designed for confidential talks through the telephone. The use of simultaneous interpreting system during the trials opened a new age in interpreting because multinational and multilingual meetings became widespread all around the world. The diplomats of those times, who were higher level aristocrats, started to appear as higher-level professionals of interpreting. Along with the development of the interpreting system facilitating multilingual interpreting, it was possible for all parties to organize such meetings, and civil society especially made good use of this equipment in their endeavour to make their voices heard. Thus, the international meetings were no longer limited to diplomatic negotiations (Baigorri-Jalón, 2014, p. 47; Doğan, 2015, pp. 10–23).

5. Protocol Rules as a Requirement of Diplomacy and Interpreting

Diplomacy and interpreting are the two professions of international environment, where certain codes of conduct called 'the protocol rules' apply. *Protocol* is a concept collocated with 'policy' and 'diplomacy' in the semantic field in the mind. Etymologically, the word 'protocol' comes from Greek *protokollon* meaning 'first sheet glued onto a manuscript', as *protos* means 'first' and *kolla* 'glue', which has then developed into 'an official record of a transaction' and finally into a general sense of 'conventional proper conduct'[8]. Protocol rules are used to maintain, arrange and provide the most appropriate medium for the statesmen and the diplomats during the conduct of international relations as well as the domestic official life and include all the codes of conduct one may need during the interaction from the order of priority, dressing to correspondence. According to Aytürk (2014, p. 7), *protocol* is the set of rules that must be obeyed during the ceremonies, formal relations, formal official correspondence and negotiations, formal visits, and meals for guests. Diplomacy, as a means of communication, encompasses a wide range of conditions including proper non-verbal communication. Criss states (2010, pp. 130–4) that in the international fora all the codes of conduct, dressing, and discourse are determined by the protocol rules; thus, the representatives paying official visits to other countries should arrange their behaviour and structure their discourse and style according to these rules, otherwise it may give rise to undesired situations that cannot be restored, ending up with political crises between the two countries.

As far as interpreting is concerned, the following are among the rules that must be considered to establish communication and interaction between the diplomatic representatives and their counterparts (Aytürk, 2014, pp. 192–3):

- In formal official matters, without the presence of other relevant people, it is not proper for a Turkish diplomat or an authority to get into bilateral communication, with the diplomat or guest authority in an isolated environment, in Turkish or in the third language without using an interpreter regardless of the fact that the guest has a good command of Turkish. However, if they are in a social environment and the foreign guest prefers to communicate in Turkish or in the third language without an interpreter, then this is considered his/her free will and preference, thus, acceptable, and natural.

8 The definition of *protocol* was retrieved from on the website of etymonline on 2.8.2016.

- In case a Turkish diplomat or authority speaks in the language of the host country in a visit to a foreign country for some business or contact provided that the contact takes place in a social environment, it gains appreciation and creates positive regard on the part of the host counterparts. In international meetings, however, the organising committee decides and announces the language(s) of the conference and it is obligatory to speak in those languages.
  An anecdote of İsmet İnönü may be thought-provoking, the brother-in-arms of Atatürk, the second President of The Republic of Turkey. He started learning English at the age of 53 in addition to French of which he had good command at the diplomatic level. In a reception, he spoke in English with the Ambassador of France and French with the Ambassador of the United Kingdom. When the reason was asked, he said, "I speak French very well but not as well as the French Ambassador and the same is true with English. I didn't want to be at a disadvantaged position just because of the use of the language."

Diplomatic interpreting is rendered in many different mediums ranging from rostrum or panel table and conference hall, round table, working lunches/dinners, receptions, visits to diplomatic, administrative, technical, or business venues. Each medium requires one or more types of interpreting service. For example, during the meeting of Joint Parliamentary Committee in Strasburg, almost all types of interpreting for the Turkish parliamentarians were needed. At the round table meeting, consecutive interpreting and sight translation of a short document were rendered; in bilateral talks carried out in a confidential medium between two persons of a crucial position, consecutive interpreting was coupled with bilateral interpreting at times; when the parliamentarians moved from one place to another, inside or outside the Parliament, escort interpreting was rendered and later when they addressed from the rostrum to the audience, simultaneous interpreting was needed. During escort interpreting it is recommended that the interpreter be alert for any sign that might start the communication at any moment from any direction. Interpreting during working lunch or dinner meetings varies according to what is need. In some small group meals, no interpreting is needed so it is possible to enjoy the whole meal as a guest at the table. However, it is probable that at times the delegation may prefer to get into communication themselves despite the fact that they do not have a good command of that particular foreign language, thus may fail to express themselves. It is recommended that the interpreter be alert about what is going on at the table to maintain the factor of 'shared knowledge' so as not to fail to mediate properly in case a situation of misunderstanding arises. In some other meals, whisper interpreting is required but then in such cases enough time is given to have meal before the meeting takes place since it would be hard for the interpreter to work during the whole course of the meal.

Press conferences or press members are another issue to handle with and should be considered within the concept of ethics. Press members sometimes try to receive some information about the meeting directly asking or indirectly – during a seemingly casual chat with the interpreter. This reality is not merely limited to diplomatic circumstances. Especially in a disaster area where there is turmoil, the interpreter comes across with these insistent demands a lot. The interpreter should make sure that it is not the task of the interpreter to provide information about the event for the news programme or newspapers. Furthermore, they may ask to put a recorder into the booth to record the interpreting. It should be noted that interpreting is a text production from one point of view and a matter of copyright; thus, recording can be accepted if they pay for it. Still, there are other factors that may have adverse consequences. Interpreting is a very different cognitive process than translating a written document. Even though all what is happening seems to be expressing what is said or written in the other language, it is well-known that a document is not a self-contained means of communication but a small portion of an extended semantic entity at least to be considered with its features expressed by the abovementioned standards of textuality. By the same token a delivered piece of speech contains much more information than what is literally said and its interpreted text can function only at the immediate moment within the experiential context with all its features such as visual or acoustic constructs that add to the semantic and pragmatic meaning of what is said. It can be misleading to listen to the interpretation later and to decipher it as if it were a reading of a written document; therefore, it must be considered from the viewpoint of pragmatics while the former from that of semantics.

6. International Relations and Diplomacy: Cooperative Principle in Discourse

International relations and diplomacy are closely associated with the principles laid on the linguistic grounds since the main tool is communication in both the oral or written sense. 'Cooperative Principle and Conversational Maxims' by Grice (1975), though not adequate to explain the complex process, draws the attention to 'cooperativeness' and to the fact that there are some principles that may help establish this quality. Doğan (2014, p. 225) refers to this principle, underscoring the importance of 'intentionality'. In every move, there is an intention. In physics, Newton's Law of Inertia states that an object will preserve its velocity and direction so long as no force in its motion's direction acts on that object. This means that if any change in the situation is targeted, then there must be a move and it is for this small step that we need to bear intention; and once the move is made in, the change starts to take place. The change can be maintained in the desired or undesired way, or in better terms, in the positive or negative way. If the target is establishing communication and interaction in agreement with the other parties, there should be an intention to come to terms, to cooperate. During daily life, one gets into contact with many people, some of whom refuse to come to terms with their interlocutors. Thus, 'establishing communication' or 'acting in cooperation' is not easy and the theoretical construct laid by Grice, comprising the Cooperative Principle and the Conversational Maxims, deserves appreciation and application.

The term 'maxim' refers to 'a general truth, fundamental principle, rules of conduct'[9]. Hatim and Mason (1990, p. 242) define it as 'a set of norms to which language users adhere in order to uphold the effectiveness and efficiency of communication'. The four maxims functioning to maintain conversational cooperativeness are those of quantity, quality, relation, and manner.

6.1. Maxim of Quantity: Interlocutors are assumed to provide as much information as required and not more than that. Furthermore, the information should be unambiguous and concise for the listener to grasp it easily. In other words, if the intention is to cooperate, then the interlocutors try to be as informative as possible. One more word, one more suffix – especially in agglutinative languages like Turkish – than what is expected or needed may end up with the violation of this maxim. If the information is more or less than what is expected by listeners, then the listeners feel that there is implicitly something more than what is to be understood easily and start to seek for more information, if not overtly, to get a better understanding.

6.2. Maxim of Quality: This maxim requires the speaker not to say what is believed to be false and for which the speaker lacks adequate evidence. The importance of 'truthfulness principle' deserves to be underlined here. The importance of truthfulness condition comes to the fore front through the experience yielded in a lifetime. When in communication with others, interpreters presuppose that what is said does not contradict with the mental constructs established in a rationale as to the world knowledge. Truthfulness is a construct more than something in the mind, which is merely true or false. It is the whole network related to a concept and the surrounding supporting constructs in the mind. The common belief is that truth is truth all over the world. However, there may be cases when truth

9 The definition was retrieved from website of Merriam Webster Dictionary on 1.1.2017.

is distorted on purpose for some reasons of benefits to the extent that the next generations come across with this distorted truth and take it as a base for the readily established network of the mind constructs. This creates a serious problem, which is hard to overcome. It is easy to cope with a lie but almost impossible to do so with the distorted mind constructs, which brings about misconceptions. Decisions made at the national and international levels affect the lives of countries and their peoples; thus, particularly the international organizations are responsible for the world peace and must act taking into consideration the human honour and dignity regardless of race, religion, ethnic origin, gender, and so forth. However, it is unfortunate that mankind come across with the violation of truthfulness condition even at the international level.

At this point an example can be given from a publication on political discourse by Schuster and Stern (2015, p. 597). A research was carried out by the participation of 50 Yale University students who conducted interviews with the Ambassadors of the USA via Skype, e-mail, over the phone, or in person. One question to the US Ambassador to Nicaragua taps the misconceptions that Nicaraguans might have of Americans and of life in the US. The following are some of the sentences from what Phyllis Powers, the Ambassador to Nicaragua, says:

> Many Nicaraguans believe that the US government and its citizens are only interested in what concerns the United States. In this region, they are thinking of security – specifically drug trafficking – and trade, which impact American businesses and American employment. That certainly is not the thing we're interested in, but that is what they think we're interested in. In Nicaragua, because of our history with the country, some think that we have imperialistic designs on Nicaragua. They do not necessarily believe just yet that we want to be partners. [...] Many still do not believe that we truly hold dear our democratic values and our desire to help others less fortunate.

This extract shows that each party has a view of life and a conception of what they are living; nevertheless, how each perceives one's life does not correspond with the other's. This is also different from what is called 'having different views'. When people have different views, they still move from the same construct in the mind. For example, if a young girl wants to go to a birthday party and the parents do not let her go there for some reasons, they argue because each side has a different viewpoint about the same reality, which is a birthday party. Their argument is whether or not it will be good for her to go there. Yet, having a different construct is not the same thing. All the postulates of logic, the universe of realities, all what is good and what is bad lay on different grounds; and even an entity in the mind does not exist in the other. For instance, one day there was a debate on TV between a fundamentalist and an expert of law. The expert was saying "...yes but what you say is not enshrined in the Constitution so you can't claim for it.".

The fundamentalist was looking at him with blank eyes because he didn't have a conception of what a Constitution was for, as he was striving for the rule of sharia.

It is, therefore, the interpreter's duty and responsibility to be alert and knowledgeable to grasp the 'universe of realities' and 'universe of discourse' of the interlocutors and skilful enough to cope with the situation of confrontation, which requires reciprocity for the Cooperative Principle to function properly for a peaceful, just, polite, constructive environment of understanding (Doğan, 2014, p. 235).

Torikai (2009, pp. 42–43) states that the quantity and quality for conversational maxims expected in a low-context culture like that of the United States cannot be applied to a high-context culture such as that of Japan, suggesting the risk of causing misunderstanding since in Japan ambiguity and vagueness are valued more than straightforward explicitness, and reflecting the high-context nature of society; the expected information given in a conversation is quite limited compared with the American way of communication. High-context culture refers to the fact that the citizens enjoy strong traditional ties in a uniform society, whereas low-context culture refers to the fact that a multinational and multicultural society like that of the USA does not entertain a collective cultural consciousness. What the interpreters do in such situations is to infer the speaker's intent judging from the context. Thus, it is the interpreter's mental and professional skills to make decisions as to how to interpret what is said by what sort of a mind for what sort of a receiver. Bulut (2015, p. 41) also underlines the fact that euphemism and metaphor as well as implicitness and indirect speech are used as strategies, instead of the daily language, by the diplomats and politicians, the experts of communication, who are capable of setting the sort of discourse that will be influential and affect opinions; thus, the same skill is also expected from the interpreter so as to attend to the same priorities and sensitivity which are put forth in the source language. Immediacy factor makes the task of the interpreter harder as the metaphor uttered by the diplomat or the politician most usually bear cultural connotations and prove to be difficult to be interpreted in the best manner.

6.3. Maxim of Relation: "The speaker is required to be relevant," is a much too simplistic description of this concept and it is not that easy most of the time. This maxim refers to the ability to identify and categorize the topic and the use of speech acts, which came into being after the linguists realized that there was more to the meaning then just dealing with it at the semantic level. Pragmatics following the emergence of pragmatism as a philosophical perspective opened new horizons so the linguists noticed that they were making a descriptive fallacy when they said, "A declarative sentence is to make an assertion or an interrogative one is to question something and an imperative is just to give orders." Thus, there came into being the speech acts, which introduced the importance of functions of the utterances. This perspective claimed that words are actions in the pragmatic sense. For example, when someone promises someone to do something saying, "I promise to do it." an act of promising is performed through words, and these are called 'performatives'. Yet, when the individual says, "I am trying to lift this bag," no word in this utterance is a 'performative' as the act of lifting is not performed through words but through psychomotor action. However, the same utterance can also imply some other thing, for example asking for help, then this act is evaluated through as a conversational implicature. This theoretical background comprises a part of the scope of discourse analysis[10].

In communication, most problems arise mainly due to two reasons:

a) The interlocutors fail to recognize and identify the functions of the illocutions uttered interchangeably during the communication process. When, for example, the main intention of speaker A is to simply express himself/herself against a situation encountered in one way or the other, he/she may do it using the discourse of an accusal, which will divert the course of the interaction towards a totally different and irrelevant direction. Then not only will the illocution be diverted but also the content of what is said will cover a different scope.

b) Sometimes the speaker can be evading consciously and on purpose for some reasons. When both interlocutors are well-intentioned to establish fruitful communication, the misunderstandings or misexpressions can be restored. When one party is in the aim to trap the other, the easiest way to do this may be the sort of discourse like evasion or coming up with a circumlocution. When the statement is terse, it is too brief, and can be much easily understood by the listener but if there is circumlocution, it may not be easy for the listener to spot and identify the problem in a short while. Hence, the relevance of the shifted or geared topic to the aim of the discourse may fail and this may or may not be a desired result. The diplomats should be and usually are the experts of language to get through such tricky discourse successfully for the sake of the good of their own countries.

6.4. Maxim of Manner: Maxim of manner demands the speaker to be brief and orderly, avoiding obscurity of expression and ambiguity in meaning. The speaker may choose to reply to a question in very short terms, which sounds curtly or tersely, thus unfriendly, and rather rude. It is preferable in a delicate medium like diplomatic environment to preserve serenity even at times when the

10 For further information, see Doğan (2014).

communication is brought to a halt. In this sense, this maxim tells us something different from what implicitness says. Implicitness is sometimes necessary when the interlocutors' intent to enable social wheels to keep turning smoothly, which is also related with the Theory of Politeness. Maxim of manner is rather about open communication in which there is no hidden agenda. Yet, it is apparent that open communication does not include the act of sharing private and sensitive information. Insight and initiative may lead a person to behave in line with what the maxims underline. Politeness Theory may, to some extent, draw the line between what is polite and communicative and what is not. The perception and application of maxims and the Politeness Theory vary from culture to culture. For example, Germans prefer direct speech, self-oriented discourse and openness; and focus on the content, whereas the British prefer indirect speech, addressee-oriented discourse, and implicitness, and focus on the audience (Doğan, 2014, pp. 270–276). Cohen (2001) dwells on the multilingualism and linguistic diversity in diplomatic contexts rather than imposing a single language to prevent confusion and asynchrony arising from the very nature of sociolinguistic aspect of languages. English provides a common denominator in negotiations; yet it is much better to resort to the multilingualism and study on the 'elucidation of lexical differences [to] overcome misunderstandings grounded in the illusion of semantic uniformity'. In conclusion, it should be noted that variety in languages and cultures would not be a source of confusion but enrichment if understanding one another's culture and language varieties were made possible through academic studies. In line with this, Karakoç (2015, pp. 64–65) states that politeness can be reflected at every level of communication from the lexical level to the discourse level and very closely related with the power relationship between the persons. Gricean maxims can shed some light into diplomatic discourse though seem to be inadequate but they may provide a means to get some taste of understanding. Usually the line between the maxims is not very definite; sometimes they overlap and sometimes fail to explain the whole communicative event; still they can help the interpreter as to what to consider.

Politeness has become a matter of academic concern in 1980s[11]. Brown and Levinson (1994) have made use of Gricean Maxims in their model of politeness, and they categorized human politeness behaviour in four strategies: *bald on record, negative politeness, positive politeness*, and *off-record-indirect* strategy. They have also said that individuals prefer to preserve face if they want to be polite but use face-threatening strategies to attack the other. Leech is another theorist who

11 For further information, see Leech (1983), and Brown and Levinson (1987).

has postulated politeness. There are eight characteristics of politeness according to Leech (2014, pp. 4–9):

a) Politeness is not obligatory so people are 'nonpolite' unless there is a reason to be polite. Though politeness is known to be a good thing, the desired one is more often thought to be impoliteness;
b) There are varying gradations of polite and impolite behaviour;
c) People behave with a sense of what is recognized as 'normal' in a society;
d) Whether politeness will occur or how far it will occur depends on the situation. Similarly, in some other situations impoliteness dominates over politeness and interestingly the reasons are more salient;
e) There is a reverse asymmetry between being polite and impolite because giving high value to the other and attributing low value to oneself is recognized as being polite; whereas giving high values to oneself and low values to the other is impolite. Leech puts forth many terms such as directness, preference or dispreference of certain behaviour, and maxims such as tact, generosity, modesty, agreement, sympathy, hedging pessimism, apologizing, indicating deference within the scope of Politeness Theory[12]. These most well-known theories concerning politeness can set the baseline and draw a frame to discuss this issue also in the diplomatic context. It seems apparent within this scope that the task of the interpreter is very challenging due to the fact that the prompt response of the interpreter is much riskier than that of the diplomat as the interpreter is mediating at least between the two cultures, two language systems, and two different mind constructs.
f) Politeness manifests itself in repetitive behaviour, which is to a lesser or greater degree ritualized. One example for this is the repetitive applause by the audience for the soloist to return to the stage to go on with the performance; another example is the diners taking turns to insist and battle to pay for the bill.
g) Politeness involves the passing of some kind of 'transaction of value' between the speaker and the other party. For example, in thanking, people thank others for something; in requesting, people make requests for some things; in making an offer, people offer something to the addressee. There are also condolences, compliments, apologies, congratulations, and so on. In this case 'reciprocal symmetry' should be considered. For example, the response to "Thank you." is "Not at all," "Don't mention it," "No problem," or alike.
h) And last but not least, politeness involves the tendency to preserve 'a balance of value' between the interlocutors. The speech acts thanking and apologizing

12 For further information, see Leech (2014).

seek to rectify the sense of debt that one interlocutor has to the other. Debt arises from the act of kindness and has to be paid back. Hospitality is repaid by a gift offered by the guest as a token of gratitude.

7. Implicitness in Diplomacy and Interpreting

Diplomatic discourse is produced by the effective use of not only language, either linguistic or non-linguistic (or extra-linguistic), but also paralanguage including prosody, intonation, stress, and so forth. It is crucial that the wording of the discourse be straightforward and as clear to be understood easily without exerting an effort as possible since all what is negotiated determines the destiny of the countries and its citizens. However, clearness is not ensured merely through straightforwardness; implicitness also requires the sort of wording that is clear to the mind of the hearer; yet, the wording is structured in such a way that it avoids sharp turns or conclusive responses so as to leave with some room for more fruitful negotiations in the quest of an agreement. In addition, politeness is an indispensable manner and style of addressing, negotiating, and expressing oneself. Implicitness, on the part of the diplomat, requires the pragmatic competence to establish a style of discourse to maintain peaceful negotiations within the scope and limits of politeness. The diplomat is expected to use the most felicitous and effective discourse to achieve the most suitable position for political action, which does not contradict with the conditions and benefits of his/her country but which also helps preserve his/her credibility as a negotiator. The interpreter, on the other hand, is responsible for establishing the most acceptable discourse to convey all what has been said and meant. Spontaneity, immediacy in the response (that is the rendition for the interpreter) is of utmost importance.

An example from the Paris Peace Conference is given below by Seymour (1965, p. 162):

> One has to be far more polite and diplomatic than in an ordinary committee. If the French make a proposition which is perfectly rotten and the chairman asks your opinion, you can't say "I don't like it," but rather: "At the present moment I feel that my government would have some hesitation in accepting the proposition of the French without reserve; may I suggest that a modification in the following sense would perhaps provide for a settlement which might, in the eyes of the inhabitants concerned, appear more equitable." I find that I am losing all capacity to say just plain "no" or "yes"; it has to be, "I should be rather slow to agree," or, "at the present moment I should feel inclined to concur."

It is said that a good diplomat is the one who never says 'no'. Negation can be made in a variety of ways. What happens in the above situation is that there is a room left to move from one view or action to another, and the final say is delayed for a later period of time to discuss and think on it. It is doubtless that this sort of a discourse is much more polite and affirmative. Implicatures, weak or strong, can be made for a fruitful and positive interaction. Another example is taken from the speeches of one of the high-level government officials, addressing the Europeans who remain indifferent against the tragedy of the Syrian refugees escaping from the war. The official person was saying:

> At a time when the repercussions of the crisis in Syria and Iraq are being felt by a number of countries, Turkey has been a representative of the international conscience by opening its doors to 2,5 million refugees and sheltering them.

The sophomores in the class, who were taking 'Discourse Analysis' course translated this sentence almost word for word as they had not taken any of the translation or interpreting courses yet. Analysis of discourse includes but it is not limited to, the Register Theory, Seven Standards of Textuality and Speech Act Theory, Propositional Analysis, Descriptive Fallacy Postulates, Cooperative Principle Maxims, Conversational Implicatures Theory and the translation theories such as Skopos, and Source and Target-oriented Approaches. Then the students came up with the result that the addressee of this speech is the Europeans, the function of the sentence (utterance) is not to assert a piece of information about the refugees but rather 'condemning', 'blaming', 'calling for action', and so forth. The first trial translations they made were unacceptable not only grammatically but also from the discourse and translation studies' viewpoints. However, the translations later were acceptable with well-formed expressions, entailing the locutions of the functions. "You are turning a deaf ear to the tragedy," "You are behaving against the human rights," and "Open your borders." were some of the locutions of the illocutions they found. More specifically, first they found the functions and the illocutions of these functions, then they transferred these functions into their locutions. The translation they made, naturally, had nothing to do with these locutions; they used the equivalence of the same lexis and syntax but the suffixes and the additional words pertaining to the language improved the translation both linguistically and pragmatically; and the interpreting rendered necessitated additional points to consider such as prosody, which refers to the natural sound and intonation. The features of diplomatic discourse were also considered and the implicitness was preserved as much as possible.

Implicitness is not an essential feature of diplomatic discourse. What makes one to prefer implicitness is the national political concerns. The diplomat speaks on behalf of the state; thus, the discourse used in the diplomatic environment does not reflect his/her own viewpoint or intention but the viewpoint of the state. Therefore, the diplomat experiences a sort of tension between the institutional personality and his/her own personality. Diplomatic discourse is the human response to the situation that has developed within a dynamic, communicative, and context-bound situation. The competence level of discourse is the hallmark of the diplomat's personality and the ultimate goal is to be able to give the most appropriate response. Sometimes an expression of an assertion with informative value is avoided because a straightforward affirmative expression or a negation or an objection can lead to undesired results as it leaves no room for safe manoeuvring. Thus, implicitness becomes crucial. According to Zhang (2014, p. 90), implicitness encompasses four terms, the semantic fields of which are different from one another:

a) *Ambiguity*
Ambiguous expressions have more than one semantically unrelated meaning and the audience is left with multiple interpretations; as is seen in *homonymy*. For example, the statement "I will meet you at the banks," creates ambiguity because 'bank' has two meanings. This sort of evasion in meaning is an unintentional one.

b) *Vagueness*
The result of the multiple readings is vagueness. *Polysemy*, is an example for vagueness: In the statement "He is good," what 'good' refers to – whether to being hardworking or trustworthy or so on – is not clear because 'good' is a subjective word so its content may change from person to person. This sort of evasion may be created intentionally.

c) *Fuzziness*
Fuzziness refers to the situation when there is referential opaqueness; for example, when the adjective 'hard-working' is used, there is no criterion to say how hard that person works. Thus, it is subject to individualistic standards of judgment.

d) *Generality*
Generality refers to the situation when an expression does not specify certain details. In other words, there is unspecification. Individuals may have different category-rankings depending on their experiences, their world knowledge, and so forth. For instance, 'my neighbour' is a general term. It does not say what sort of a person that individual is, male or female; young or elder, and so on.

These expressions demand more efforts to be made at both the intralingual and interlingual levels. Thus, the diplomat and the interpreter should be competent in the use of source and target languages, respectively.

Diplomatic correspondence is usually different from diplomatic oral discourse since the written material requires 'entailment'; whereas oral discourse mostly makes use of 'implicatures', whether weak or strong. There are many more academic articles and books published on the written mode of diplomatic discourse. Here are some of the references: Aksoy (2001); Bhatia (2006); Bulut (2015); Di Carlo (2015); İskit (2012); Karakoç (2015); Şentürk (1995); Tezok (2014); Vural (2007). In addition, Doğan (2014, pp. 155–6) elucidates the difference between the oral text in a speech and written text under the title 'Mode of Discourse', which is illustrated by Gregory and Carroll (1978, p. 49). They classified only the oral or written texts and the clues for interpreting can be drawn from what has been postulated by them. According to Gregory and Carroll, 'mode' refers to the medium of the language activity and there are 'various permutations' on the distinction between a written text and a spoken text. First of all, speaking can mainly be of two types: *spontaneous* and *non-spontaneous*. Spontaneous speech can be of two types: *conversing* and *monologing*. These sorts of speeches do not pose much challenge during the course of interpreting because spontaneity brings forth natural sound, that is prosody, which makes 'meaning assembly' in the mind easy. On the other hand, *non-spontaneous speech* is made either by reciting or by speaking of what is written. From the interpreting perspective, the real challenge starts at this point. Speakers make use of a written material in different ways, some are helpful for the interpreter, some are not, even for the audience listening to it in the same language. However, the speaker may speak extemporaneously; it is noteworthy that not all spontaneous speeches are extemporaneous or vice versa. For example, TED Talks are extemporaneous but not spontaneous because the speaker gets prepared for them in advance, which poses a risk for the interpreter because the speeches in that programme are usually coupled with a sense of stage show. In a spontaneous or extemporaneous speech if the speaker uses extensive cultural elements, then interpreting can be very difficult and probably ends up with losses if the speaker has not cooperated with the interpreter before the event. Otherwise, this sort of speech is a very easy one to interpret because it is not highly structured in the syntactic sense like a text written to be read silently and the lexical density is usually low. According to Gregory and Carroll, the speaker may make use of the written material in various ways. They state that in case there is a written material before the delivery, the speech made by a speaker can be 'spoken as if not written', 'written to be spoken', or 'written not necessarily to be spoken', and even this last one has two variations. Baranyai (2011, p. 8), who wrote about interpreting in diplomatic contexts, also states that "The aim of translation and interpretation is not the semantic and syntactic transmission of the sentences from one language to another, but to achieve the same effect with the translation as what the original version is supposed to achieve." Besides, Doğan (1996) lists the challenges an interpreter faces during the rendition: the ones stemming from the text written, from the speaker making the speech, or from the time constraint. Interpreting of each delivery type bears in itself many challenges, necessitating the strategies to cope with. Few speakers are conscious of the difference between a text prepared to be spoken or read, and the variations between them. Yet, it is the interpreter's or a translator's task to know how to treat any sort of a text and what to do while receiving it as an input and reproducing it in the other language as well as making use of certain signs around to maintain equivocality, and getting feedback. Non-verbal information and codes of conduct in a diplomatic environment are the other things to be considered. To elucidate some points described here, an anecdote of Doğan may well serve as an example, which is about the speech of one of the ex-ministers of Turkey. When editing a book having been translated from a Turkish, there was the printed and translated version of a speech. It is important to note that an orally presented material, in other words a speech, can be transliterated to be published in a book but the degree of this transliteration differs in terms of the sort of material to be published, the purpose of its publication, and the target population to whom it addresses. There are many other factors to be considered. Sometimes it is proper to transliterate without changing anything but sometimes appropriation is needed for the transliterated text to function in the desired way. In this experience, unfortunately, the register, mainly the mode of discourse, was not considered by the translator. In the original speech, the minister in a small conference hall complains to a certain group of people working in that area about another sector drawing attention to some of the deficiencies, and utters an adjective exacerbatingly chosen to mean 'terrible'. However, it is very well-known that what he says does not really apply to the whole system. There is nothing wrong when such a delivered speech and its word for word translation is published on a webpage, in a book for a particular sector, or for all the citizens, but if it is to be translated and published in a book being prepared for the foreigners, the word 'terrible' is not the right translation because that word was uttered as if having blurted out due to sincerity; a euphemistic expression which is not that sharp should be used for the transliterated text to serve its purpose. It can be a statement like "This sector has been in the development process," which still conveys the idea that there is something lacking. This sort of translation or interpreting is not a word for word one but can be called a 'pragmatic' one. A research was carried out by Zhan (2012, p. 214) about the roles of the interpreters in diplomatic contexts and the results demonstrated that even the staff interpreters working for the government and interpreting in political settings do not always act in strict accordance with the professional norms and rules, and prefer to act as mediators of communication especially when they are rendering dialogue interpreting. Hence, sociocultural differences; political sensitivities and reserves; political facts and events as well as diplomatic maneuvres carried out so far; historical interaction between the states and the future they are aiming at; and the roles and responsibilities of the interpreter are all very important. Therefore, the training of translation and interpreting of diplomatic contacts should be a separate course in the curriculum.

8. Educational Implications and Concluding Remarks

Training of translation or interpreting of diplomatic discourse is a process including a wide range of topics to be learnt and skills to be developed. The practices in the past may well shed light into the present and future practices, consolidate the decision-making processes of the interpreters, and strengthen the immediate response skills during the interaction. Discourse Analysis and Critical Discourse Analysis enable the students to see behind the eyes, hear behind the utterances and read between the lines. Analysing strong and weak implicatures and entailment as well as the use of body language from the speaker's point of view; ostension (in other words, how to make use of the body language of the speaker) from the hearer's point of view, which are covered by the Relevance Theory; and socio-psycholinguistic analysis making use of Politeness Theory, are all included in such a training. Interpreting, by itself, is specifically challenging due to the immediacy and simultaneity factors so thinking in the quick-minded manner is a skill to develop. It is vital that not only the target population but also the intention in the source text or speech, which lay the grounds for a particular expression, and the textual traditions of both language systems as well as functions of the illocutions be considered in addition to the mode of discourse which refers to whether it is spoken or written, written to be spoken or expressed in the target language to be read or listened to, and so on. Practices to interpret emphasizing different focus points such as theme or rheme, message or intention should be provided. In addition, Codes of Conduct in a diplomatic environment that the interpreter should obey and display is another training objective. Diplomatic discourse patterns should be well grasped and their immediate reproduction in the target language should be maintained. At this point Translation Studies will enlighten the complicated issue of what to say and how to say in the target language, keeping in mind the sociocultural features and ethical values, and rules. Training of implicitness in both source language and target language and reflecting the shades of meaning of the content in the target language is another issue to dwell upon. Restatement exercises may well pave the way towards better expressions especially if reciprocity in the interaction is concerned. In situ and in vivo practices should accompany in vitro ones carried out in the isolated atmosphere of the interpreting laboratory or class-room. The interpreter should have a wide and deep knowledge of diplomatic and political background but this may not be sufficient; being aware of the personal discursive preferences of the interlocutors in interaction during the diplomatic contacts may facilitate the interpreting process. Furthermore, training on creativity, sense of humour, and handling with the culture-bound discourse preferences of the interlocutors as well as the idiosyncratic expressions may all make up the curriculum for interpreting in diplomatic contexts. The roles and responsibilities of interpreters, positioning oneself in the environment and what to/not to expect from the organizers of the event, or to what extent are also of crucial importance.

In conclusion, to be a diplomat is very difficult and stressful but being the interpreter of a diplomat is much harder and more stressful because the diplomat after all is saying what he/she has in his/her own mind within his/her universe of realities and discourse; yet, the interpreter interprets what is in the mind of the diplomats conversing or negotiating in any manner. To be able to 'position oneself' in the diplomatic environment may help mental skills function much more freely and even creatively for the diplomats while negotiating and for the interpreters while interpreting. What should be kept in mind that before text there is context, before gist there is intention and before tactics there is strategy. Let's hope for a better future where there is diplomatic politeness to preserve peaceful interaction and well-trained interpreters accordingly all over the world.

References

Abu Jaber Kamel S. (2001). Language and Diplomacy. In J. Kurbila and H. Slavik (Eds.). *Language and Diplomacy* (pp. 49–54). Malta: DiploProjects.

Aksoy, Berrin N. (2001). Siyaset Dili ve Çevirisi. [Political Language and Its Translation] Durak M. (Yay.Haz.) [Editor]. *Kuramsal ve Uygulamalı Çeviri Sorunları [Problems Concerning Theoretical and Practical Translation]* (pp. 139–145). 3–4 Kasım 2000 Sempozyum Bildirileri [3–4 November 2000 Symposium Presentations]. Ankara: TÖMER Yayınları [TÖMER Publications].

Alexander, Philip. (2014). The Rabbis, the Greek Bible and Hellenism. In J. K. Aitken and J. Carleton (Eds.). *The Jewish-Greek Tradition in Antiquity and the Byzantine Empire*. Cambridge: Cambridge University Press.

Aytürk, Nihat. (2014). *Protokol Yönetimi: Kamusal Yaşamda Protokol Kuralları. [Protocol Management: Protocol Rules in Public Life]* (7. Baskı) [7th ed.]. Ankara: Türkiye ve Ortadoğu Amme İdaresi Enstitüsü [Public Adiminstration Institute for Turkey and Middle East].

Baigorri-Jalón, Jesús. (2014). From Top Hat to Short Sleeves Diplomacy: Two Centuries of Diplomatic Interpreting. *UN Chronicle*. No: 3.

Baranyai, Tamas. (2011). The Role of Translation and Interpretation in Diplomatic Communication. *Journal of Translation and Interpretation* (pp. 2–12). Vol. 5. No: 2.

Barston, R.P. (1997). *Modern Diplomacy*. (2nd ed.). Essex: Longman.

Brown, Penelope. and Levinson, Stephan. (1994). *Politeness: Some Universals in Language Usage*. (2nd ed.). (1987, 1st ed.). Cambridge: Cambridge University Press.

Bulut, Alev K. (2015). *Tercüme Hatası!?* [*Translation Error!?*]. Istanbul: Çeviribilim Yayınları [Çeviribilim Publications].

Cáceres-Wursig, Ingrid. (2012). The Jeunes de Langues in the Eighteenth Century: Spain's First Diplomatic Interpreters on the European Model. (pp. 127–144). *Interpreting*. Vol. 14, No: 2, Amsterdam: John Benjamins Publishing Co.

Cohen, Raymond. (2001). Language and Negotiation: A Middle East Lexicon. In J. Kurbila, H. Slavik (Editors). *Language and Diplomacy*. (pp. 67–92). Malta: DiploProjects.

Cremona, Vicki Ann and Mallia, Helena. (2001). *Interpretation and Diplomacy*. In J. Kurbila and H. Slavik (Editors). *Language and Diplomacy*. (pp. 301–305). Malta: DiploProjects.

Criss, Nur Bilge. (2010). Komşularla Sıfır Sorun Çok Güzel Bir Söylem ama Gerçekçi Değil. ['Zero-Problem with Neighbours' Is a Good but Unrealistic Discourse] Özdal H., Dinçer OB. and Yeğin M. (Yay. Haz.) [Editors] *Mülakatlarla Türk Dış Politikası* [*Turkish Foreign Policies with Interviews*]. Cilt 3. (pp. 130–134). Ankara: Uluslararası Stratejik Araştırmalar Kurumu (USAK) Yayınları [International Strategic Research Organisation Publications].

De Beaugrande Robert and Dressler, Wolfgang. (1981). *Introduction to Text Linguistics*. London: Longman.

Di Carlo Giuseppina, Scotto. (2015). "Weasel Words" in Legal and Diplomatic Discourse: Vague Nouns and Phrases in UN Resolutions Delating to Second Gulf War. *International Journal of Semiotic Law*. Vol: 28. (pp. 559–576).

Diriker, Ebru. (2015). On the Evolution of the Interpreting Profession in Turkey: From the Dragomans to the 21st century. In Ş. Tahir Gürçağlar, S. Paker, and J. Milton (Eds.). *Tradition, Tension and Translation in Turkey* (pp. 89–107). Amsterdam: John Benjamins Publishing Company.

Doğan, Aymil. (1996). Yazılı Metinden Sözlü Çeviri ve İlgili Eğitim Programı İçin Bazı Öneriler [Implications for Sight Translation and Its Training]. *Çeviribilim ve Uygulamaları* [*Journal of Translation and Interpreting Studies*] (pp. 25–35). Ankara: Bizim Büro Basımevi [Bizim Büro Publishing Company].

Doğan, Aymil. (2000). İspanya'da Bir Çeviri Konferansının Ardından [In the Aftermath of a Translation and Interpretation Conference in Spain]. *Mediterraneo.* No: 2. İstanbul: Çantay Basımevi [Çantay Publishing Company].

Doğan, Aymil. (2012). Modernism Movements through Translation and Interpretation Activities in Ancient Anatolia, Ottoman Empire and Republic of Turkey. In R. Kumar (Ed.). *Role of Translation in Nation Building* (pp. 57–73). New Delhi: Modlingua.

Doğan, Aymil. (2014). *Sözlü ve Yazılı Çeviri Odaklı Söylem Çözümlemesi (Interpreting and Translation Centered Discourse Analysis).* Ankara: Siyasal Yayınevi Siyasal Publishing House].

Doğan, Aymil. (2015). *Sözlü Çeviri: Çalışmaları ve Uygulamaları* [*Interpreting: Studies and Practices*]. (4th ed. Updated and Expanded Version/1st ed.-2003). Ankara: Siyasal Publishing House.

Gregory, M., Caroll, S. (1978). Language and Situation: Language Varieties and Their Social Context. London: Routhledge and Kegan Paul. In B. Hatim, I. Mason (1990) *Discourse and the Translator.* London and New York: Longman.

Grice, H. Paul. (1975). Logic and Conversation in P. Cole, J.L. Morgan (Eds.), *Speech Acts* (pp. 41–58). New York: Academic Press.

Hatim, B. and Mason, I. (1990). *Discourse and the Translator.* London: Longman.

Hitzel, Frédéric (Ed). (1995). Enfants de langue et Drogmans [Dil Oğlanları ve Tercümanlar]. Mehmet Sert (Translator). Istanbul: Yapı Kredi Yayınları [Yapı Kredi Publishing].

Hoffman, Joel M. (2004). *In the Beginning: A Short History of the Hebrew Language.* New York, London: New York University.

Hurford, J.R. and Heasley, B. (1987). *Semantics: A Coursebook* (pp. 15–30). Cambridge: Cambridge University Press.

İskit, Temel. (2012). Bozatlı, B. (Yay. Haz.) [Editor] *Diplomasi: Tarihi, Teorisi, Kurumları ve Uygulaması* [*Diplomacy: History, Theory and Practice*]. İstanbul: Bilgi Üniversitesi Yayınları [Bilgi University Publications].

Jönsson, Christer and Hall, Martin. (2002). Communication: An Essential Aspect of Diplomacy. *43rd Annual ISA Convention, New Orleans, LA. Panel WA23: Diplomatic Theory and Practice.*

Karakoç, Nihal Yetkin. (2015). *Çeviri ve Diplomasi* [*Translation and Diplomacy*]. İstanbul: Çeviribilim Yayınları [Çeviribilim Publications].

Leech, Geoffrey. (1983). *Principles of Pragmatics.* London, New York: Longman Group Ltd.

Leech, Geoffrey. (2014). *The Pragmatics of Politeness.* Oxford University Press. (1st published in 1983 by Longman).

Matteucci, Aldo. (2001). Language and Diplomacy: A Practitioner's View. In J. Kurbila and H. Slavik (Eds.), *Language and Diplomacy.* (pp. 55–65) Malta: DiploProjects.

Onat, Şefik. (2016). *Bir de Bizden Dinleyin* [This Time Listen from Us]. İstanbul: Alfa Yayıncılık [Alfa Publishing].

Paker, Saliha. (2009). Turkish Tradition. In M. Baker and G. Saldanha (Editors). *Routledge Encyclopedia of Translation Studies* (pp. 550–559). Oxford: Routledge Publication Company.

Pöchhacker F. (2004). *Introducing Interpreting Studies.* New York: Routledge.

Searl John R. (1983). *Intentionality: An Essay in the Philosophy of Mind.* Cambridge: Cambridge University Press.

Setton Robin. (1998). Meaning Assembly in Simultaneous Interpretation. *Interpreting: International Journal of Research and Practice in Interpreting.* (pp. 163–99), 3:2. Amsterdam: John Benjamins.

Seymour, Charles. (1965). *Letters from the Paris Peace Conference.* New Haven: Yale University Press. (Quoted in Baigorri-Jalón Jesús. 2014:146).

Şentürk, Tahir. (1995). *Courses on Diplomatic Correspondence.* Ankara: Dışişleri Eğitim Merkezi [Foreign Affairs Training Center].

Schuster, Justin and Stern, Eric. (2015). Interview with Francis Riccardone, Jr by Fiona Lowenstein. *Diplomatic Discourse.* Cambridge: Cambridge University Press.

Tezok, Ahmet Funda. (2014). *Diplomacy and Diplomatic Correspondence.* Ankara: Hacettepe University Publications.

Thiéry, Christopher. (2015). Diplomatic Interpreting. *Routledge Encyclopedia of Interpreting Studies* (pp. 107–108).

Timur Ağildere, Suna. (2010). The Importance of Teaching Turkish as A Foreign Language in Europe in the 18th Century: Ecole des Enfants de Langue d'Istanbul in the Ottoman Empire (1669–1873) *Turkish Studies.* Vol. 5, Issue:3. (pp. 693–704).

Torikai, Tumiko. (2009). *Voices of the Invisible Presence.* Amsterdam: John Benjamins Publishing Company.

Vural, Sergül Kara. (2007). Diplomasi Metinleri Çevirisi Örneğinde Kaynak Metin – Erek Metin Bağıntısı [Source Language and Target Langue Relation in Diplomatic Text Translation]. *Mersin Üniversitesi Dil ve Edebiyat Dergisi* [Mersin University Journal of Linguistics and Literature]. 4:1. (pp. 49–67).

Zhan Cheng. (2012). Mediation through personal pronoun shifts in Dialogue Interpreting of Political Meetings. *Interpreting.* Vol. 14. No: 2. (pp. 192–216). Amsterdam: John Benjamins Publishing Co.

Zhang, Junfeng. (2014). From Interpersonal to International: Two Types of Translation in the Making of Implicitness in Diplomatic Discourse. *Perspectives: Studies in Translatology*. Vol: 22. No: 1. (pp. 75–95). Taylor&Francis Group, Routledge.

Electronic Sources

Bhatia Aditi. (2006). Critical Discourse Analysis of Political Press Conferences. *Discourse & Society*. London, Thousand Oaks, CA and New Delhi: SAGE Publications 17:2. (pp. 173–203).
http://das.sagepub.com Vol 17(2): 173–203)f1

Diplomacy
www.etymonline.com/index.php?allowed_in_frame=0&search=diplomatic

Diplomatic crisis
http://www.haaretz.com/news/peres-humiliation-of-turkey-envoy-does-not-reflect-israel-s-diplomacy-1.261381

Diplomatic and Consular Relations

a. Vienna Convention on Diplomatic Relations (VCDR)
http://www.mfa.gov.tr/data/kutuphane/multilateralconventions/viennaconventionondiplomaticrelations.pdf
b. Vienna Convention on Consular Relations (VCCR)
http://www.mfa.gov.tr/data/kutuphane/multilateralconventions/viennaconventiononconsularrelations.pdf
c. Guide to Diplomatic Missions in Turkey
http://www.mfa.gov.tr/data/Guide-to-Diplomatic-Missions-in-Turkey.pdf

Dragoman
www.liquisearch.com/dragoman/history

Interpreter
www.etymonline.com/index.php?allowed_in_frame=0&search=interpreter

Maxim
https://www.merriam-webster.com/dictionary/maxim

Policy
www.etymonline.com/index.php?allowed_in_frame=0&search=policy

Protocol
www.dictionary.com/browse/protocol

Assist. Prof. İnönü Korkmaz Ph.D.

Trakya University
Department of Translation and Interpreting

Translation of Technical Documentation for the User of Electrical and/or Electronic Goods

Abstract: Technical documentation of the user of electrical / electronic goods and their translations are of significance in today's world of technology. From the translation studies' point of view, the text types should be dealt with in the very first place. The distinction between the 'scientific' and 'technical' texts and their translations are accordingly addressed. To support the base of such a distinction, Byrne's (2012) categorization of scientific and technical texts is consulted. Pursuant to this categorization, user-oriented documentation – specifically the manuals and reference documents – is explained as a subcategory. In addition, text typology (Haramundanis, 1992) and text analysis (Eruz, 2008) within the perspective of translation studies are also referred. More specifically, documentation consisting of guides and manuals are explained. The translation of user-oriented technical documents is touched upon and the concepts of technical translation or translation of specific field texts are viewed as well. Besides, standardization of technical documents for user products is another key concept to be examined. Thus, international and national standardization bodies, such as ISO, and directives (73/23/EEC EU Directive) are included within this context. Finally, communicating and translating technical documents are exemplified with an effort to inspect how technical documents are constructed and which translation strategies are used in the translation process of such texts. Since, a great number of the user-oriented technical documents consist of user guides and user manuals, the examples are provided from such texts. In this regard, a quick view on translators' decisions and prospective views on the issue of technical translation are also addressed.

1. Introductory Remarks

Living in a world where almost each day a new piece of information is rapidly processed and transforms into something concrete, only inches away our fingers can make us feel that we will not be able to catch up with these novelties one day. It is true that we are living in a very fast-paced world and in every part of our lives, we can witness the speed of technological developments. Only a few decades ago, we were amazed to see holographic three-dimensional messages rising on a small piece of gadget on cult science-fiction movie scenes and dreamt of having one of these at home. Moreover, some of us even felt a little bit scared to see robots doing much of the work such as constructing other machines, doing housework or even doing surgical operations since this would mean unemployment for most of us and finally, the extinction of humans and emergence of the world of machines. Actually, the real scenario does not necessarily have to be so pessimistic should you put the technological developments in the military products apart. As human beings, we have the ability to use the information which we gather from many diverse disciplines to our best advantage and to make new products which can ease our lives. These products may vary from smart phones to mobile stroke treatment units equipped with Computerised Tomography (CT) scanners. The art of combining different disciplines and thus, bringing out so many products is one thing, the art of presenting all these to us, to human beings with different educational and cultural backgrounds, and physical and mental states, or many other differences likewise, is another. In other words, inventing a machine is undoubtedly a difficult task; however, helping people to use it correctly is not easier at all. At this stage, documents or texts designed to help people understand the operation of these machinery, or to give people instructions to operate them, or briefly to benefit from these come into prominence for users of these technological products. In addition, owing to the global trade connections, the translation process of these texts has to be considered with similar weight of care. In order to reveal the significance of these texts and their translations, in the present part of this book, an insight to the documents related to some of the technological products, from the eyes of a user keeping the translation process in mind, is provided. The examples of the documents and most of the standards and directives related to these documents have been chosen from those of Turkish so as to give a profile of the Turkish / English and in reverse direction of translation process for technical documentation.

In terms of translation studies, there have been many attempts so far to distinguish different text types. Before addressing these attempts, it might be more fruitful to look back at the ideas founding the basis for such text type diversifications. In the second quarter of the previous century, Karl Bühler (1934/1984, p. 35) came up with the idea that language had three main functions: *expression*, *representation*, and *appeal*. This concept clarifies the semantic perspective of language and gives way to further discussions about the functions of texts and consequently to the classifications of texts. In 1970's, Katharina Reiss adopted Bühler's classification for functions of language in search for a better understanding of the concept of '*equivalence*' in translation studies. She favoured a different route in addressing the issue of '*equivalence*' by viewing the *text* rather than *word* or *sentence* as opposed to her antecedents (Reiss, 1981). Through this view and with the base of Bühler's classification upon language functions, Reiss (1971/2000) mentioned about three different text types: ***Informative*** texts, ***Expressive*** texts, and ***Operative*** texts. She further added a fourth type of ***Audio – medial*** texts. Informative texts aim at plainly communicating facts and thus, could be used to convey information, knowledge, and opinions. Expressive texts aim at expressing the attitude of the sender of the message and deal more with the aesthetic form of the message. The main goal of the operative texts is to encourage the receiver of the message to operate in the way that the sender wishes and by doing so, the text is produced for the purpose of stimulating the receiver to accomplish the desired response. Finally, audio-medial texts, which Reiss added in this classification later, could fall into the category of films and visual-audial advertisements that could be seen as supplementary to the previous texts' functions. This categorization of text types might be useful for us to navigate through text types and find a suitable place for the technical documentation. Among the mentioned types of texts, the informative texts could be seen as the most prominent candidate associated with the technical documentation. If informative texts can be classified as the texts aiming to convey different types of information, knowledge, and facts on a particular subject, so can the technical documentation.

Now it could be assumed that the technical documents might fall into the category of informative text types in Reiss' classification. However, there is still more to discover in a sole technical document than just simple signs of information dissemination. First of all, a simple text might have more than one aim as observed in the case of a technical brochure. To give an example, a cellular phone producer company might need to ***express*** their attitude to their potential customers in order to get a bigger share in a new market. Bearing this in mind, they might need to re-design one of the products' technical documents, namely a user guide, in a way that the target audience will benefit from it and recommend the product to the prospective users. In order to do this in the process of writing and translating, which could be defined as re-writing of the source text in the target language, developing a creative human skill might be needed. This skill might be nourished from both the writer and translator's professional knowledge as well as that of linguistics. As Douglas Robinson (1997, p. 149) defines, translators and interpreters usually make their living by pretending someone else who are experts in a field that the translators and interpreters have actually never been. They get into these characters and use their creative and linguistic skills in order to '***fake***' the audiences. Paul Kussmaul (1995, p. 33) defines this skill as 'expert behaviour' and calls for it as '*required*' for the sake of a successful translation.

Klaus Gomlich (in Wright, S.E. and Wright, L.S. Jr., 1993) draws the attention to another classification of the technical texts: ***transfactual*** and ***transbehavioral.***

The ***transfactual*** type deals with the enhancement of the reader's knowledge based upon a specific field. Since this could be seen as a general description, Gomlich adds two sub categories as ***transfactual I*** and ***transfactual II. Transfactual I*** type of texts aim at the audience equipped with some level of expert knowledge. For example, a scientific presentation submitted in a conference where the audience has already been familiarised with the topic. On the other hand, ***transfactual II*** type of texts aim at the target audience with non-expert knowledge.

The ***transbehavioral*** texts could be defined as the texts that attempt to influence the reader's behaviour. This definition is also divided into two sub-categories as ***transbehavioral I*** and ***transbehavioral II***. The first one comprises a larger variety of texts from personal letters to the advertisements. The behaviour of the recipient of the texts could be altered or influenced but this is done without any fixed set of rules, laws or directives. On the other hand, ***transbehavioral II*** type of texts aim at fulfilling this by the routines of society, imposed laws, standards or manuals.

Some technical documents might integrate more than one feature and show different writing styles than the others. A manual for a smart phone, for instance, might include parts which aim at not only behaviour change in the recipient but also giving information on different levels since the users' age and experience level vary from a teenager to a veteran citizen.

Another dimension of a technical documentation is that it has its own style and form. Yet, there seems to be varying views in establishing one fixed style or form. In a study of technical texts and the dogmas on the same issue, Karen Korning Zethsen (1999, p. 66)[1] typifies the syntactic features of technical texts such as:

- *nominalisation,*
- *heavy pre- and post-modifications,*
- *use of third person, and*
- *long and complex sentences.*

Mark Herman (1993) also makes a contribution by giving a full description of technical translation and focuses on three main features of technical texts: ***clarity***, ***concision*** and ***correctness***. According to Herman, the yield of a technical translation process, namely the translated text, should be clear enough to convey the knowledge easily (clarity), short enough to avoid the reader becoming frustrated

1 Retrieved from the website of Journal of Linguistics, Aarhus University on 8.12.2013.

(concision), and finally correct enough to hinder readers colliding against the facts of the expertise field (correctness).

As seen above, the endeavour of defining technical documents and understanding their nature is a challenging one. The first and utmost reason for this challenge could emerge from the bias upon defining what is technical and what is not. As a translation scholar and translator Jody Byrne (2006, p. 7) criticises that there is still a mistake of using terms ***scientific*** and ***technical*** interchangeably. He admits that these terms are so-close but also points to the fact that they actually denote to very distinct notions. The term *scientific* deals with the accumulation of the facts and studying them for the sake of human beings while the latter focuses on the issue of how to process the information gained by science and bring it into people's use. To get a better grip of the concept, it is feasible to investigate through the general technical document types and see if it is possible to draw an outline on the concept of technical documentation on the premises of translation studies.

2. Technical Documentation Types

Before scrutinising the technical documentation types, it would be more appropriate to grasp the concept of '*technical*' from the point of translation. As Byrne (2006) pointed out, there is a blur distinction between the terms '*technical*' and '*scientific*' when commenting on the text types. The terms are generally used together and in some concepts, they are used interchangeably in a mistaken manner. Earlier, in a study of technical and scientific translation, Isadore Pinchuck (1977, p. 13 in Byrne, 2012, p. 2) underlines the key categories of information that can provide the materials for scientific and technical translation as:

- results of pure science,
- results of applied scientific studies carried out in order to solve a specific problem, and
- work of technologists with the aim of presenting an industrial product or service intended for sale.

Pinchuck clearly states that there is always an overlap between these categories and this leads to the fact that modern scientists' theoretical works today are ready to be converted into tomorrow's technological products. Even though there is a close relation between the bases of information or knowledge in technical and scientific texts, the way how information is presented and used will undoubtedly differ. While technical texts are designed to communicate the information as clearly as possible, scientific texts are designed to discuss, analyse, and synthesize the similar type of information (Byrne, 2012).

In his book, Byrne focuses on the technical documents and states that technical writers deal with a great variety of documents in very distinct subject areas from gardening tools to children's toys. Despite the fact that his book mainly deals with software documentation or more specifically software user guides, Byrne borrows the classification of technical documents in these headings from Mike Markel:

- proposals
- reports
- instructions

In this classification, according to Markel (2012, p. 440) *proposals* are the documents which are designed to propose an idea as a basis for a product or service before it is launched on the market. Whether it is planned for a piece of cosmetics' clinical trial by a pharmacological company or an outline for toys' safety instructions by a toy producing company, the main aim is to persuade the reader to accept the basic terms and conditions. Since they can integrate financial information, legal issues, and technical engineering materials together, *proposals* could be challenging for both technical writers and translators.

As Byrne (2012, p. 51) points out, *reports* are designed to give detailed information relating to a product, namely, *its development stages, viability, commercial success, safety, effectiveness, and so on*. Reports are divided into three main categories: informational, analytical, and recommendation reports.

As for the *instructions*, Markel believes that they are the most common type of technical documents that are designed to help readers to fulfil a task. They might include repair manuals, administration guides, and user guides. Whether it is a user manual or a user guide, one of the most common characteristics of these documents is that they are intended primarily to ensure the safety of the user and to prevent any damage to persons or their properties.

Byrne (2012, pp. 58–69) clearly distinguishes technical and scientific texts and asserts a more comprehensive classification as well. His list seems rather an integrated one:

a) *Manuals*
This group includes cookbooks, tutorials, guides, reference manuals, and even software such as online help pages.

b) *Applications and Proposals*
This group could be described as both scientific and technical. For instance, a proposal for an academic research project might include information on its duration, budget, materials, project inputs and outputs, and so on, and these

all need to be designed in a way that the text could be defined as both technical and scientific.

c) *Reports and Scientific Papers*
These are also a type of hybrid texts like proposals and applications. For instance, a scientific paper could be used to exchange information among scientists working in a particular discipline using the special language and terminology of this discipline.

d) *Presentations*
Even though they are not acknowledged so, PowerPoint presentations could also be identified as specific type of texts.

e) *Regulatory Documents*
They could also be named as normative documents such as national and international standards and directives such as ISO standards and EU directives.

f) *Popular Science Texts*
These texts aim at educating common people on novel scientific and technical information.

Another classification comes from Katherine Haramundanis (1992), who is also interested in the computer industry. She also uses three main headings in order to differentiate between the document types by uncovering the main purposes of the texts such as *marketing materials*, *materials that report* and finally *instructive materials* as shown in Table 1. Documents enumerated under the marketing heading are designed to convince or persuade readers, whereas the reporting materials aim to state the facts related to the product. Lastly, materials under the instructing heading are intended to both instruct and train the readers.

Table 1. Technical Documentation Types

Marketing	**Reporting**	**Instructing**
brochure	magazine article	customer manual
case study	newspaper article	user manual
sales pamphlet	journal article	instruction manual
press release	internal publication	site preparation manual
product handbook	technical paper	installation manual
product catalogue	progress report	owner's manual
marketing script	internal report	reference manual
marketing talk	annual report	maintenance manual
sales presentation	blueprint	system manager's manual
technical summary		operator's manual
advertising copy		technical description
white paper		functional specification

Marketing	Reporting	Instructing
'mock' paper testimonial data sheet product brief application guide		user interface specification glossary training manual quick-start guide presentations course materials

(Haramundanis, 1992, p. 3)

As clearly seen from these classifications, technical documents are either employed to inform readers about a service or product's ongoing development starting from its designing stage to the selling stage, or to persuade readers to accept a particular idea, or to instruct or train the users of the products or the beneficiaries of the services. For instance, an *owner's manual* is designed to help the user to perform basic functions of the product. Or, a *user interface specification* could help a computer user to understand the newly designed interface of a software in order to benefit from the software to its best. In other words, they can inform us that *x* machine has *y* number of features or they can be used to fulfil a particular task or service and finally guide us how to make it work safely and effectively. Among these aims of technical documents, the instructing and guiding ones are significant for the users of a product. It is now feasible to deal with the types of technical documents of electrical/electronic goods, specifically intended for the users.

3. User-Oriented Technical Documents

In the context of technical documentation for electrical and/or electronic goods, users generally benefit only few of the aforementioned classified texts. These might be gathered under the title of *instructive* texts. Although Haramundanis' list covers the products related to the computer industry, it is notable that the third group of instructive documents in her categorization could also be defined as a neighbouring one with those of the electrical and/or electronic goods. Especially the items such as *instruction manual, user manual, installation manual, owner's manual, reference manual* and *quick-start guide* (see Table 1) could also be considered as applicable for electrical and/or electronic goods. Besides, apart from the software products, all other computer products such as hardware, pc output devices, hard disks, external memory devices, and so forth would undoubtedly be counted as electrical and/or electronic goods as well. These instructing documents designed for the operation of such goods are often defined as *manuals* or *guides*. Then, it is also necessary to define a manual or a guide in terms of technical writing.

A manual or a guide is a piece of technical document designed to describe a product or service in detail in order to help the user understand how it works. Moreover, a manual or a guide aims to instruct or train the user for a better utilisation of the product in addition to providing a safe usage. In a study of scientific and technical writing styles, Philip Rubens (2001, pp. 13–17) provides a variety of manual types keeping an eye on the user needs:

- Tutorial guides / training guides
- User guides / operating manuals
- Reference manuals
- User reference manuals
- Job aids / quick-reference guides

User guides and/or operating manuals are the most common technical document types intended for the users of a product. As Rubens (Ibid, p. 15) points out user guides and/or operating manuals show users the specific step-by-step procedures and instructions necessary to use the product purchased. The distinction between reference manuals and user reference manuals might be noted as the fact that reference manuals could aim at the readers of a greater variety such as both users of a product or the craftsmen engaged in the servicing of the same product. Both a text containing information on how to drive a car safely and comfortably and a text on how to change the engine of the same car can be considered as *reference manuals*. However; *user reference manuals* aim at only the readers who could be the users of the related product. Through a close examination of such products and their documentation, there seems to be another variety of definitions used to name these documents:

- Quick Start Guide
- Installation Guide / Manual
- Quick Installation Guide / Manual
- Easy Guide
- User's Guide / Manual
- User Guide / Manual
- E-manual[2]

The reason for such a variety clearly results from the contents that the related documents cover. If a document accommodates for a list of installation instructions,

---

2 This classification is obtained by the examination of the sample documents and used for evaluation process of the doctorate dissertation by Korkmaz İ. (2016).

it is inevitable to call it an '*installation guide*'. Similarly, if the document covers information on functions of a product on a basic level, it could be labelled as '*quick start guide*' or '*easy guide*'. Most of the time, the names of the documents involve 'quick' or 'easy' when they contain short and basic information. For commercial or other reasons, many producers are inclined to provide their customers with electronically created documents, namely 'e-manuals', which are not published on paper, but rather created in the electronic environment and given to the users in CD formats or on the website. Apart from the others, 'user guides' and/or 'user manuals' are generally presented in the longer format texts including a *contents* page, and mainly dealing with sections such as *general specifications*, *safety precautions*, *installation instructions* (when required) on *how to use / operate*, *using instructions*, *possible troubles* and *troubleshooting*, *service information*, *warranty* and *liabilities*. Even though these sections may not be applicable for a specific product or goods, most of the documents for electrical and/or electronic goods need to cover them according to national and international standards or national directives and regulations, which will be discussed in detail in the section concerning the standardization of documents.

So far, how technical documents could be defined and separated from the other forms of texts have been explained – starting from the text typology in translation studies to the awareness of how scientific and technical texts may be addressed apart from each other. Furthermore, the separation of user oriented technical documents from other technical documentation has revealed that the user guides or user manuals are the predominant technical documents for user products.

There also lies the concern of reaching a standard in the translation process. Even within the same translation agency, one translator might follow a different route than the others. Or worse, a company's products might be traded in a country with a wrong advertisement/marketing policy without noticing the users' cultural differences and the target text can end up as a source of insult. In addition, to make it worse, there might be cases of injuries and damages emerging from the misunderstandings of the users about the functions of a product while using it. To prevent such incidences, both technical writing and translating processes, or at least their outcomes should reach a standard level. The efforts aiming at this standardization will be dealt in the following section.

4. Standardization of Technical Documents of User Products

It is undoubtedly a difficult task to standardise even a slightly important issue in the modern world of technological improvements. One of the main reasons for this is the fact that a vast variety of areas may be collaboratively working together in order to yield a product. Take the example of a cellular phone: in order to produce a cellular phone, experts from such fields as electronics, mechanics, communication, software engineering, ergonomics, and so forth should be commissioned to collaborate. In the context of technical documentation writing and translating, users' needs, experiences, and knowledge level; sociological and cultural factors; safety and environmental issues, and so on must be taken into consideration. With the help of standards comprising organisations (such as ISO) and national and international boards of directives and regulations on trade (such as Turkish Standard Institute (TSE) and EU Directives), this task would become relatively easy. In an international scope, one of the most commonly referred organisations is International Standard Institution, usually known with the abbreviated form of ISO. Founded in London in 1946, ISO mainly covers the mechanical engineering issues and is an independent and non-governmental international organization having representatives from nearly every country in the world.

One of the most commonly consulted ISO standards[3] relating to the user oriented technical documentation is ISO/IEC GUIDE 37 which was drafted in 2009 and is now in use. It has the title of '*Instructions for Use of Products by Consumers*' and covers the main titles of *design* and *communication*, *legibility of texts*, *colour*, *communication principles*, *wording* and *use of technical terms*, *language(s)*, *illustrations*, *graphical symbols*, *tables*, *flow charts*, *table of contents*, *action in the event of errors* or *malfunctions*, *warning notices*, *durability* of *instructions for use*, *evaluation*, and *checklists* for the instructions in the documents[4]. This standard is translated into many languages and have become applicable for the national standards of the related countries. For instance, GB 5296.1-2012 of National Standard of the People's Republic of China; IS/ISO/IEC Guide 37: 2009 of Indian Standard with the same title and TSE ISO/IEC GUIDE 37: 1999[5] of Turkish Standard Institute (TSE) are in line with the ISO standards addressing to the same issues. Before moving to the national scale, it is significant to refer to the commission for electrical and electronic technologies throughout the world. As seen in the ISO standard name (IO/IEC GUIDE 37), International Electrotechnical Commission

3 A *standard* is usually a technical document which provides requirements, specifications, guidelines or characteristics of materials, products, processes, and services in order to ensure that they all fit their purpose. The definition was retrieved from the official website of ISO on 15.5.2016.

4 DRAFT ISO/IEC GUIDE 37, revision of the second edition of ISO/IEC GUIDE 37:1995, contents page.

5 This is also the converted form of an earlier ISO standard.

(IEC) is another liable international body for the standardisation and conformity assessment missions.

In a much narrower scope, the European Parliament enacted the DIRECTIVE 2006/95/EC,[6] binding among the European Union member countries, so as to harmonise the laws concerning electrical equipment designed for use within certain voltage limits. The directive mainly deals with the legal issues on safety precautions, conformity of the equipment, and general conditions relating the products. Operating in Turkey, TSE issued a standard, TS 11823, relating the guides for consumer products in 1995. This standard covers the rules for introductory and user guides for the traded goods in Turkey, and consists of two main sections. In the first section are general rules on legibility of texts, contents of texts, using instructions of the product, safety precautions, warranty and liability issues, and environmental issues regarding the disposal of the products. The second section[7] gives out the specific rules on the storage, consignment, installation, maintenance procedures, service providers, and expected lifetime of the products.

Apart from the standards, regulations on the issues of trading and free circulation of goods are periodically issued in order to keep a solid norm for such practices. Issued in 2006, 73/23/AT[8] is a revised version of an earlier regulation and a similar design of the 73/23/EEC EU directive. It simply deals with safety precautions, standards referred, CE conformity marking, technical dossier relating the documents, market surveillance, and measures.

It is obvious that both in national and international scopes, there is an escalated interest in the attempts of standardisation of the technical documentation, oriented for the users of electrical and/or electronic goods. With the help of the regulations, directives, and rules[9] emerging from the efforts of these bodies,

---

6 DIRECTIVE 2006/95/EC (2006): *Directive 2006/95/EC of the European Parliament and of the Council of 12 December 2006 on the Harmonisation of the Laws of Member States Relating to Electrical Equipment Designed for Use Within Certain Voltage Limits* (codified version). Official Journal of the European Union, L 374/10-18, issued on 27.12.2006.

7 TS 11823, *Guideline for Identifying and Using Rules*, Turkish Standard, issued on 25.09.1995.

8 Regulation relating the electrical goods designed for use within certain voltages. Turkish Republic Official Gazette, Issue 26392, dated 30.12.2006.

9 A *regulation* is an act of legislation and it is binding. Therefore, it must be applied in its entirety in all the EU countries. A *directive* is also a legislative act which sets out a goal that all EU countries must achieve in a certain period of time, yet it is left to the individual countries' preference to devise their own laws on how to reach these goals. A *guide* is usually the devised version of the directives on national scale and published

technical documents of traded goods would become uniform and beneficial for both the producers and the customers of the related products. Besides, the communication between these two parties would become more efficient and fruitful since the technical documentation is the most powerful key to compromise with each side. Thus, how this communication is realized via either technical documents or their translations will be the topic of the next section.

5. Translation Process of Technical Documents

An efficient way of writing technical documentation starts with a thorough analysis of the audience. In the case of technical documents for user-oriented products, this audience becomes the users of the products. Therefore, after determining the target audience, the first goal of a technical documentation is considered to be satisfying the users' needs. This needs analysis might show us how to treat user documentation. In a study on writing 'usable user documentation', Edmond H. Weiss (1991, pp. 10–11) compares user documentation with works of art and concludes that the user documentation is in fact a *device* which could aid the communication between technical writers and users. According to him, a well written literary text might need to meet the criteria of beauty, elegance, and high class. When a text is seen as a device of communication between technical writers and their readers, then the criteria of maintainability and reliability, which are borrowed from the engineering fields, need to be met. In the process of writing user documentation, specifications should be tested and refined, and the view of the reader should be considered. Another view supporting this idea comes from Byrne (2006, p. 99): for technical communicators, the primary objective is to make user guides more accessible and usable. As he points out, *usability* means focusing on users. Therefore, it is vital for a technical writer to understand the audience at the beginning. Similarly, in the training process, a prototypical model could be created and the translation trainee should be convinced that he/she needs to know why he/she is translating a specific piece of technical document and who will be the target text reader and for what purposes that specific piece of technical document will be used.

When it comes to the translation of these documents, is it possible to follow the same claims? With the escalation of functionalistic approaches (owing to Reiss and Vermeer), the equivalence concept has been dealt with on the text level and translators try to look behind the functions of both source text and target text. In

in the Official Gazette of the related countries. These explanations were retrieved from the EU official site, http://europa.eu_on 15.5.2016.

a way, understanding a source text's function and its prospective readers via the needs analysis of this new audience might help a translator to produce the target text to be understood by its readers. It is certain that understanding the needs of the audience will not be sufficient on its own since the source text itself might have not been understood by the translator at the beginning.

Thus, for the translator, the very first step is to embrace the source text. In translation studies, this has its own place as a translation oriented text analysis. Also, in translation training, this analysis would open up the route for translation trainers in their search on how to prepare their trainees for the professional work environments.

In the Turkish tradition, Sakine Eruz (2008, pp. 57–59) points out the significance of translation oriented text analysis. In a didactic context, she draws an outline consisting of four steps in order to teach translator trainees how to analyse a source text before moving into the translation phase:

a) It is essential to improve the trainees' comprehension skill and to test it in order to make them understand the source text.
b) The trainees should base their understanding which will be tested by the tutor.
c) The trainees should create a text with their own words using the source text as a starting point.
d) The trainees should adjust the knowledge they gained to the other texts (genuine expression within the context of communication).

These steps are considered as pre-translation work and should become an automatic process for further professional situations. Having worked as an experienced translator / interpreter in courts of Germany, Eruz (2000, pp. 99–114) also emphasizes the importance of using *parallel texts* within the context of translation of specific field texts. The use of parallel texts in translation of technical documentation could be helpful especially for the user manuals or guides since they represent similarities both in their form and style. In other words, a sample of a user manual or a guide in the target text could be resorted as a parallel text. Bearing this in mind, in the process of technical translation training, the trainees should be encouraged to have insight of the different types of technical documents. They should also be given the chance of examining the related parallel texts. By doing so, they would compare their translations with these parallel texts and would be able to produce more acceptable translations.

After the pre-translation work, there might still be situations where different factors could hinder the translation process. Cultural differences between the source and target audiences, differences emerging from socioeconomic status, age, and gender could represent such barriers. Yet, with the help of translation strategies, these obstacles might be removed. As Vinay and Darbelnet's (in Jeremy Munday, 2012, pp. 85–89) classification of *direct* and *oblique* translation strategies might be a resort for technical translators as well. Under the title of direct strategies are the procedures of *borrowing*, *calque*, and *literal translation*, while the other procedures of *transposition*, *modulation*, *equivalence*, and *adaptation* fall under the oblique translation strategies (these strategies will be exemplified later). It is certain that these strategies are not solely intended for the use of translators of technical documents and may not be helpful in every challenge that the translator faces. The key for conveying ideas and information from the source text to the target text successfully is rather more complicated. In fact, a translator's work is beyond conveying ideas and information. There is the risk of getting into the habit of stitching a few tested and succeeded strategies and overlooking the minor issues and missing the bull's eye. In the following section, the translation process of these texts will be mentioned in order to gain an insight into the technical documentation on the basis of translation studies.

6. Strategies Pursued in Technical Document Translation

At this point, it is necessary to admit that the evaluation of translation products is a risky process. One might easily fall into the trap of hunting for errors in the translated texts. Before stepping forward a little further, it is important to emphasize that the aim here is not to criticize the translators or their exemplified works. The main objective here is to gain an insight how the translation process is performed and what results might emerge in the end. The sample texts are selected from Turkish-English / English-Turkish translations just to open a window to the novel translation trends for technical documents in Turkey.

In an effort to create coherence, it is essential to review the translation strategies and to present a few related examples. It might be useful to remind the definitions of each strategy, which possibly facilitates to see how these strategies work between the Turkish and English language pair and to evaluate the results at first hand.

6.1. Direct Translation Strategies

Among the direct translation strategies are borrowing, calque, and literal translation.

6.1.1. Borrowing: Transferring a source language (SL) unit (mostly at one-word level) directly to the target language (TL) is in fact, the most common strategy in the technical and scientific translations between English and Turkish since many technical and scientific terms and words relating to a specific field of science in modern Turkish are borrowed from foreign languages.

Example 1

SL word	TL word	Reference
Flash	***Flaş***	*Fuji finepix S6000fd* digital camera owner's manual, p. 15.
Wireless ***adaptor***	*Kablosuz* ***adaptor***	*Inca*, Wireless USB adaptor User Manual, p. 10.
Scart cable	***Skart Kablo***	*Next HC12* Digital HD Cable TV Receiver User Manual, p. 7.

6.1.2. Calque: It is also a kind of borrowing, yet this time the transfer is done in a literal translation mode. Although there could be a semantic change in some cases, borrowings and calques could become integrated in the TL. In this strategy, SL units are transferred into TL units through a process in which units already existing in the TL are reshaped according to their meanings in the SL.

Example 2

SL word	TL word	Reference
Extension cord	*Uzatma kablosu*	*Fakir DARKY 1600* Stick Vacuum Cleaner Instruction Manual, p. 4.
Rubber collar	*Lastik manşet*	*BOSCH WMV 1610* Washing Machine User Manual, p. 13.
Emniyet Termostatı *	*Safety Thermostat*	*Arzum AR 010* Electrical Instant Water Heater User Manual, p. 4.

* the word 'termostat' is actually a borrowing in Turkish, but since the translation direction is from Turkish to English this could be considered as back translation.

6.1.3. Literal Translation: It is the word-for-word translation strategy and is more common between the language pairs sharing similar language family and having similar syntactic structures. Thus, it is not observed very widely between the language pairs of Turkish and English since both languages belong to different language families.

As it is mentioned before, the main objective here is to gain an insight how the translation process is performed and what results might emerge in the end.

All the direct translation strategies could be considered to be fitting in with the word-for-word translation concept.

6.2. Oblique Translation Strategies

In the event that direct translation strategies result in an unacceptable target language unit or a disagreement among the authorities, then the oblique translation strategies need to be applied. For example, the English word 'selfie', which means 'taking a photo of oneself using a smart phone'. Even though this word could be commonly used in Turkey, Türk Dil Kurumu (TDK) [Turkish Language Society][10] opened up a contest for its equivalent in Turkish and as a result, the word 'özçekim' which is a combination of 'öz' (meaning 'self') and 'çekim' (meaning 'shooting a photo') was coined.

Transposition, modulation, equivalence, and adaption strategies are the examples of the oblique translation strategies performed in technical translations.

6.2.1. Transposition: This could be defined as a change in the part of speech of language unit, such as converting a verb into a noun or vice versa. This strategy is called upon when there is no possibility of using a literal translation strategy.

Example 3

SL word	TL word	Reference
... cihazınız kullanma kılavuzundakinden ***farklı*** *olabilir... (adjective)*	*... your refrigerator may* ***differ*** *slightly from ... (verb)*	*Simfer BZ 2507* Refrigerator User Manual, p. 3.
... for ***emergency calls*** *... (noun)*	... ***acil arama yapmak*** *için* .... (verb)	*LG G2* Smart Phone, Quick Start Guide p. 9.

6.2.2. Modulation: This could be defined as the semantic or structural changes between SL and TL. Changing a passive voice structured sentence into an active voice structured sentence (or vice versa) could provide a good example for modulation.

Example 4

SL word	TL word	Reference
... ***carry out*** *the first wash cycle without any laundry ... (active voice and imperative)*	*... ilk yıkama işlemi çamaşırsız yapılmalıdır... (passive voice – not imperative)*	*BOSCH WMV 1610* Washing Machine User Manual, p. 9.

10 TDK is an official society founded in Turkey with the aim of conducting research on the Turkish language and dealing with the related issues. This explanation was retrieved from the website of TDK on 18.5.2016.

SL word	TL word	Reference
... ***are** the batteries **drained**?* ... *(passive voice – present tense)*	*...piller* **bitmiş** *mi?* .... *(active voice – past tense)*	*Samsung HT-D350* Digital Home Entertainment System, User Manual p. 29.

6.2.3. Equivalence: According to Vinay and Darbelnet (1995/2004), this term is used to refer the cases where different languages might describe the same situation with different styles or structures. It is more common in the translation of proverbs and idioms, so it is not very common to find examples in the technical documentation. Nevertheless, when the technical brochures or user manuals serve as advertising tools as well, companies might choose to use certain mottos (as an example, the German car producer Audi's reputed 'Vorsprung durch technik' finds its English equivalent as 'Progress through technology') in order to attract the consumers.

6.2.4. Adaptation: This is the change of a cultural reference when the situation does not exist in the target culture. Even though it is more common in the literary texts or texts related to arts, a contemporary term seems to be replacing it; namely localization. However, it might not be described as making a small unit of language familiar in the target culture, rather it could be best described to make the whole text or documentation familiar in the target culture.

Even though there are other supplementary translation techniques in Vinay and Darbelnet's study such as *amplification*, *false friend*, *explication*, and *generalization*, the most common translation strategies and procedures have been touched upon. Nonetheless, as seen in the examples given above, it cannot be asserted that all the strategies are equally applied in translation of user-oriented technical documents.

7. Concluding Remarks

To sum up, it is significant to make a clear distinction between the scientific texts and those of technical even though the background knowledge for both seems quite similar. After distinguishing these two texts, it is useful to be aware of the subcategories of the technical documents or texts. Due to the text typological attempts and also the functional approaches in the field of translation studies, making a map of technical documents has become an easier task.

As previously mentioned, user-oriented technical documents generally consist of user manuals, user guides, and instructive manuals. They are designed to instruct the users on how to use a product or service and help them to use the product safely and accordingly. In order to achieve these objectives, they need to be functional, informative, and instructive as well as being capable of meeting the needs of users. The translations of these documents should also aim at the same targets. To produce a user document or its translation more effective, it is essential to fulfil the necessities of product users, official standards, directives, and regulations on both national and international scales. In a way, these standards, regulations, and directives have become the manuals or guides for the preparation of such user documents and thus, they could be considered as technical documents as well.

Finally, the translation of user-oriented technical documents is a demanding job since it requires not only the technical knowledge and good linguistic skills but also an enthusiasm for search on user profiles; cultural differences; socio-economic status gaps between the prospective users; age and gender oriented differences; needs analysis of the users and producers; and so forth. Besides, translation-oriented text analysis and when needed, ability to find parallel texts and being able to benefit from these texts could be counted as other merits of a translator. Therefore, the training on technical translation should be aimed at instructing the translator trainees on how to harmonise their linguistic skills with the technical information in special fields along with good research skills.

References

Bühler, Karl. (1984). *Theory of Language: The Representational Function of Language.* Donald Fraser Goodwin (translator). Amsterdam and Philadelphia: John Benjamins Publishing Company.

Byrne, Jody. (2006). *Technical Translation: Usability Strategies for Translating Technical Documentation.* Dorrecht: Springer.

Byrne, Jody. (2012). *Scientific and Technical Translation Explained.* Manchester: St Jerome Publishing.

Eruz, Sakine. (2000). Çeviride ve Çeviri Eğitiminde Koşut Metinler. [Parallel Texts Used in Translation and Translation Training] *Alman Dili ve Edebiyatı Dergisi* [Journal of German Language and Literature]. No: 4248. İstanbul: İstanbul Üniversitesi Rektörlük Yayını [Published by Istanbul University Rectorate].

Eruz, Sakine. (2008). *Akademik Çeviri Eğitimi: Çeviri Amaçlı Metin Çözümlemesi.* [Academic Translation Training: Translation-Oriented Textual Analysis]. İstanbul: Multilingual.

Gomlich, Klaus. (1993). Text Typology and Translation-Oriented Text Analysis. In Sue Ellen Wright and Leland D. Wright, Jr. (Editors.) *Scientific and Technical Translation.* American Translators Association Scholarly Monograph Series.

Volume VI (pp. 175–184). Amsterdam and Philadelphia: John Benjamins Publishing Company.

Haramundanis, Katherine. (1992). *The Art of Technical Documentation*. Massachusetts: Digital Press.

Herman, Mark. (1993). Technical Translation Style: Clarity, Concision, Correctness. In Wright, Sue Ellen and Wright, Leland D. Jr.(Eds.). *Scientific and Technical Translation*. American Translators Association Scholarly Monograph Series. Volume VI, (pp. 11–19). Amsterdam and Philadelphia: John Benjamins Publishing Company.

Korkmaz, İ. (2016). *Teknik Metinlerde İşlevsellik: Kullanma Kılavuzlarının Çevirileri* [*Functionality in Technical Texts: Translations of User Manuals*]. Doctorate Thesis, İstanbul Üniversitesi, Sosyal Bilimler Enstitüsü, Çeviribilim Anabilim Dalı [Istanbul University, Institute of Social Sciences, Department of Translation Studies]. İstanbul.

Kussmaul, Paul. (1995). *Training the Translator*. Amsterdam and Philadelphia: John Benjamins.

Markel, Mike. (2012). *Technical Communication*. (10th ed.) (pp. 431–578). Boston: Bedford/St. Martin's.

Munday, Jeremy. (2012). *Introducing Translation Strategies: Theories and Applications*. (3rd ed.). New York: Routledge.

Pinchuck, Isadore. (1977). *Scientific and Technical Translation*. London: André Deutsch Ltd. In Byrne, Jody. *Scientific and Technical Translation Explained*. (2012). Manchester: St. Jerome Publishing.

Reiss, Katharina. (2000). *Translation Criticism: Potential and Limitations*. E.F Rhodes (Translator). Manchester: St Jerome and American Bible Society.

Reiss, Katharina. (1981). *Type Kind and Individuality of Text: Decision Making in Translation. Poetics Today*, Vol. 2, No. 4.

Robinson, Douglas. (1997). *Becoming A Translator*. (pp. 148–152) London and NewYork: Routledge.

Rubens, Philip. (2001). *Science and Technical Writing*. (2nd ed.). London: Routledge.

Vinay, J. P. and Darbelnet, J. (1995). *A Methodology for Translation*. In Lawrence Venuti (Editor). (2004). *The Translation Studies Reader*. (2nd ed.). London and New York: Routledge.

Weiss, Edmund H. (1991). *How to Write Usable User Documentation*. (2nd ed.). Phoenix; Oryx Press.

Electronic Sources

Regulations, Directives
http://europa.eu/eu-law/decision-making/legal-acts/index_en.htm

Standards
https://www.iso.org/about-us.html

Türk Dil Kurumu (TDK)
http://www.tdk.gov.tr/index.php?option=com_content&view=article&id=77&Itemid=1

Zethsen, Karen Korning. (1999). The Dogmas of Technical Translation – Are They Still Valid?. *Hermes – Journal of Linguistics*. Volume 23. Denmark: Aarhus School of Business, Aarhus University.
http://pure.au.dk//portal/files/9952/H23_05.pdf

Assist. Prof. Zeynep Oral Ph.D.

Hacettepe University
Department of Translation and Interpreting

Translated by Halise Gülmüş M.A.

Turkish Sign Language and Its Applications[1]

Abstract: Sign languages are the visual-spatial languages that are employed by the hearing impaired to communicate with each other and with those who can hear. The common denominator of sign languages are gestures, facial expressions, the finger alphabet, and the position of the body and hands. Contrary to the popular belief, sign languages are not universal and the sign language of each country reflects their culture, their world view, and discourse of the society to which the hearing-impaired belong. Turkish Sign Language (TID) differs from other sign languages with its own grammar, vocabulary, phonetics, and forms of expression. After being acknowledged that sign languages were complete languages like spoken languages, the field of sign linguistics have gained momentum and has become the subject of research in translation. When the European Parliament acknowledged sign language as an official language in 1988, sign language interpretation began to be accepted as part of the interpreting studies. In Turkey, the importance given to Turkish Sign Language (TİD) has increased in line with the non-disabled society policy aiming at the incorporation of the deaf community to the society as well as keeping them informed about the world events and developments. Thus, the need for sign language interpreters multiplied. In Turkey sign language interpreters are generally the hearing children of deaf adults known as CODA (child of deaf adult). CODAs voluntarily or involuntarily act as a bridge in communication from an early age. Due to the fact that sign language was banned for years, its vocabulary remained limited compared to the vocabulary of Turkish language. Thus, the interpreters have to deal with some problems during the interpreting process. According to the Theory of Sense developed by Seleskovitch and Lederer, a successful interpreting is to get the sense/meaning of the language as well as the thoughts and feelings of the speakers and signers express.

1 The part is translated and developed from Zeynep ORAL's article 'Türk İşaret Dilinde Karşılığı Olmayan Sözcüklerin Çevirisinde Uygulanan Bazı Stratejiler [Some Strategies Employed In The Translation of Non-equivalent Word In Turkish Sign Language] which was published in EFD / JFL Journal of Faculty of Letters Volume 32/Number 2/ December 2015.

1. Definition and Features of Sign Languages

Signs have been practiced and evolved into language systems since the beginning of human communication. According to American psychologist Merlin Donald (1993, p. 740), "Mimetic action is basically a talent for using the whole body as a communication device, for translating event perceptions into action". He asserts that mimetic thought is the ability of Homo Erectus to produce conscious non-linguistic representational acts. Clearly, signing has been practised since ancient times. Gestures, mimes, and imitations were the things that were used for communication at the time of Homo Erectus and not only spoken languages but also sign languages were constructed on these.

Hearing and Deaf[2] people live in the same society and share the same cultural identity; however, Deaf people have developed their sub-culture and sign language is an indispensable part of this Deaf sub-culture and a primary means of communication for Deaf people. Sign languages, which enable the deaf to communicate with each other and with the hearing ones, have their own grammatical structures, sign systems, and syntax. Turkish philosopher Nermi Uygur (2013) defines 'native language' as the language of the community in which a person was born and grow up. With reference to this definition sign languages can be assumed to be native languages of the deaf. Sign languages are virtually visual-spatial languages; thus, they generally use manual and non-manual means to establish communication including hand signals, movements of hands, finger alphabets, grammatical facial expressions, and body movements. However, sign languages vary from one region to another. For instance, while American Sign Language and French Sign Language are one-handed languages, Turkish Sign Language use the two-handed manual alphabet. Hand shape (configuration), movement, location, and orientation are the features of sign languages.

Until the 1960s sign languages were seen as primitive languages that are limited or inadequate for proper communication (Açan Aydın, 2013, p. 75). Because of this, sign languages weren't used in educational institutions, which naturally brought about the fact that hearing-impaired people were socially excluded from the society for years. However, sign languages have been one of the major concerns of linguistics since the 1960s. The first ground-breaking work in this field is *Sign Language Structure* by Stokoe (1960). This publication was followed in 1965 by *A Dictionary of American Sign Language on Linguistic Principles*, which is the first dictionary of ASL by Stokoe and his two deaf colleagues (Stokoe, Casterline, and

2 The first letter of the word 'Deaf' is capitalized herein so as to emphasize both the hearing-impaired people and their culture.

Croneberg, 1965). Studies show that sign languages are not inferior to spoken languages; they fulfil the same linguistic, social, and cognitive functions as the spoken ones (Stanler and Martin, 2006 in Makaroğlu, 2012, p. 6). Having been approved that sign languages are also full languages like their spoken counterparts; linguistic, cognitive, and structural studies on sign languages have gained momentum and since those times they have been among the most valued research topics within the scope of translation and interpreting studies.

Sign languages are considered as natural languages, yet they are different from spoken languages in terms of communication channels, grammatical structures, and phonetics. According to Cuxac (1993), there are five differences between the sign languages and their spoken counterparts: configuration (hand shape), orientation (rotation of the hand), location (place of the hand), movement (motion of arm, wrist, fingers and hand), and facial expressions.

According to these five factors, the sign for the verb 'living' is as shown in Figure 1:

Figure 1. 'Living' in Turkish Sign Language

- Configuration : the fingers are straight and adjacent to each other
- Orientation : the right hand is horizontal, the left hand is vertical
- Location : the left hand is on the chest level, the right hand on the face level
- Movement : asymmetrical, synchronized, while the left hand moves up vertically, the right hand moves from the wrist to the elbow horizontally
- Facial Expression : no mimics, while the sign for the verb 'living' is shown, the word is pronounced with the lips as well

As can be understood from the instructions and seen in the picture, the verb 'living' is written as a single word; however, in the sign language, this word can be expressed in a three dimensional way.

While in spoken languages sound is produced by vocal fold vibration thanks to small muscles, in sign languages a sign is produced with the help of muscles of the hand and arm and it takes more time (Moody et al., 1998, p. 75). Besides hands, facial expressions, and body movements are also important parts of sign language production. According to Dikyuva et al. (2015), all of these features of sign languages constitute its grammar, which includes lexicology, phonology, morphology, and syntax. Drawing an object, measuring distance, specifying the size or position of an object, describing movements, directions, time, and even colours are possible in sign languages (Delaporte, 2005, p. 44).

2. Sign Language Interpreting as a Communication Tool

European Commission defines sign language interpreter as a professional who transcodes spoken language into sign language or vice versa. The interpreters are supposed to facilitate communication and ensure equal access to information and participation. Hearing children that grow up in a deaf household are called as 'CODA' in English, which is an acronym standing for Children of Deaf Adults. In Turkish, these children are also called as 'KODA'. However, some linguists prefer 'SEÇ' as an acronym which stands for 'Sağır Ebeveynlerin Çocukları' instead of 'KODA'. In Turkey KODAs, who are virtually bicultural and bilingual since they acquire spoken language and sign language at the same time, are generally assumed to be the interpreters of their parents.

As previously stated, sign languages are different from spoken languages in terms of their grammatical structures. Therefore, Turkish Sign Language is also different from spoken Turkish. Sign languages not only have the properties of natural languages such as double articulation, arbitrariness, mobility, change, cultural transmission, but also Jacobson's six functions of languages: referential, emotive, conative, phatic, metalingual, and poetic functions (Açan Aydın, 2013, p. 75). All of these features enable interpreting between sign languages and their spoken counterparts.

Although channels are different in sign language interpreting, the main purpose is to convey the messages between two languages. There are some studies assessing different interpreting models on sign language interpreting. These studies show that sign language interpreters' mental process and that of spoken languages interpreters are not different from each other (Pointurier 2014, Ricono 2004).

Jacobson (1963) outlined the linguistic model of interpersonal communication and mentioned about six significant elements of communication.

Figure 2. Jacobson's Communication Model

Context (reference)

Addresser ------------ Message -------------Addressee

Channel

Code

These six elements are fundamental components of communication. If the addresser and the addressee do not use the same channel and language, these are barriers to their communication. The following table elucidates the comparison between sign languages and spoken languages in terms of the six elements of communication:

Table 1. Comparison between Spoken and Sign Languages in Terms of the Six Elements of Communication

	Spoken Language	**Sign Language**
Addresser	A hearing and speaking person	A person who is hearing-impaired or uses a sign language
Addressee	A group of people who can speak and hear	People who are hearing-impaired or know sign language
Context	Reference	Reference
Channel	Voice	Facial expressions, gestures, fingers, body position
Code	Spoken language	Sign language
Message	Verbal	Visual

In this communication model, translators or interpreters only intervene in the 'message' part of the communication because the main aim of the translators or interpreters is to convey the message objectively and properly to the target audience. In *simultaneous interpreting*, the speaker is in the meeting room and the interpreter works in a soundproofed booth. While the speaker is speaking into a microphone, the interpreter receives the sound through earphones and conveys the message into the target language via a microphone almost at the same time. Generally, the audience does not see the interpreter in this kind of interpreting.

However, in *consecutive interpreting* the interpreter waits for the speaker to finish one or two sentences and following this, the interpreter conveys the messages to the target language. When it comes to sign language interpreters, they translate messages between the sign and spoken languages. Sign language interpreters may sit or stand, but they should position themselves in the best place to be seen by the deaf audience because in sign language interpretation the 'channel' is the interpreter together with his/her body positions and gestures. Thus, the sign language interpreter is always visible. Apart from this, in simultaneous and consecutive interpreting the 'channel' is the 'voice' of the interpreter. All of these differences between sign and spoken language interpreters result from their way of rendering messages to the target languages.

The expressions, word choices, and linguistic level that are proposed by the interpreter during interpreting, represent the character and style of the interpreter. However, in sign language interpreting mimics and dialect differences give clues about the translator's personality, culture, and the universe of discourse. As might be seen, messages can be rendered in many ways as long as the meaning is conveyed properly. This situation is the same in both sign and spoken languages and it underlines 'langue' and 'parole'.

3. The Interpretive Theory of Translation

The Interpretive Theory of Translation (the Theory of Sense) was developed by Danica Seleskovitch and Marianne Lederer in the 1970s. According to this theory, interpreting is a process in which the meaning of the source utterance is transferred into the target language with the aim of producing the same response action in the target audience. This process is divided into three stages: comprehension, deverbalization, and reformulation. Seleskovitch and Lederer (2002) clarify that as the main aim of the interpreting is not the transfer of words, but conveying the meaning, interpreting is not performed at language level but at speech level.

While translating from a sign language into a spoken language or vice versa, it is not possible to translate each word with a sign. As stated before, syntax and semantic structures of sign languages differ from their spoken counterparts. Therefore, the interpreter who doesn't know how to transfer a word from spoken language into a sign may face some problems in the interpreting process.

4. Sign Language Interpreting

Interpreting processes from Turkish into TSL and from TSL into Turkish will be examined within the framework of the Theory of Sense, developed by Seleskovitch and Lederer.

4.1. Simultaneous Interpreting from Turkish to TSL

Phonemes are produced successively in spoken languages; nevertheless, they occur simultaneously in sign languages. In the sign language interpreting process, the interpreter should use both physical features such as facial expressions, hand and body positions, and mental ones at the same time. They also switch the medium of communication from auditory to visual-spatial during interpreting.

Simultaneous interpreting from Turkish into TSL can be scrutinized in eight stages:

4.1.1. Hearing the Message

This stage is not only about hearing but also focusing on the message of the speaker and listening to it carefully to get the meaning properly.

4.1.2. Understanding and Analysing the Message

This stage requires a long attention span. The interpreter's ability to understand the speaker's intention and the implied meaning of the message and then rendering this intended meaning via the most suitable signs; all these are directly related to the interpreter's competence in synthesizing.

4.1.3. Comprehension Process

At this stage, as previously mentioned, the meaningful units of the source utterance are to be kept in mind. The interpreter may accomplish a successful interpreting as being aware of the differences between 'langue' and 'parole'.

4.1.4. Memory and Organization Processes

This stage is called as 'deverbalisation'. With reference to 'The Interpretive Theory of Translation,' deverbalisation is the process of focusing the attention on the intended meaning rather than the words in the original language. Deverbalisation has a vital role in the process between the comprehension of the utterance and its reformulation in the target language. Interpreters are expected to understand the emotional and cognitive meanings apart from the linguistic signs. Thanks to the translator's cognitive memory, deverbalisation is virtually practised (Lederer, 1994, p. 23).

The example of deverbalization: "Sayısal lotodan iki kolon oyna" which means "Play the lottery by selecting numbers from the first two columns" in English, can be translated into TSL as "lottery, two different parts, choose six numbers." If the interpreter renders word for word translation in that kind of situation, the hearing-impaired person may not understand the meaning of the message properly since the Turkish word 'kolon' does not have any equivalence in TSL in that context. To be able to render the meaning, the interpreter ought to be competent at analysing and synthesizing. Apart from knowing TSL, qualified interpreters should immerse themselves in the deaf community to understand their culture, language, and lifestyle. Hence, if the interpreter only translates each word with a sign, it is not interpreting but it is just transcoding Turkish words into signs.

4.1.5. Finalizing the Process

Even though comprehension and re-formulation of the message occur simultaneously, the interpreter's comprehension and reformulation process should go through maximum 6 seconds' time lag (décalage), depending on the interpreter and the difficulty of the message (Ricono, 2004, p. 24). Seleskovitch and Lederer (2002) define this time lag as 'the time difference between comprehension and reformulation processes.'

4.1.6. Evaluating the Quality of the Interpreting

Interpreters ought to evaluate the quality of their interpretation at the end of the process. In general, they confront some problems while translating from spoken Turkish into TSL. This is caused by some vocabulary of Turkish which do not have any equivalent sign in TSL. As opposed to popular belief, the vocabulary of TSL is not limited and the hearing-impaired people do not have any communication problems resulting from the vocabulary. The challenges which interpreters encounter during the process result in poor quality interpretations, the decrease in the pace and effectiveness of the interpretation, and problems about rendering the exact meaning. Therefore, if there is no equivalence of a Turkish word in TSL, there may be semantic losses on the message and impact levels.

The strategies used by certified sign language interpreters to solve a problem they have encountered while interpreting from Turkish to TSL can be listed as follows:

- *Fingerspelling*
  Explaining the problematic word by using the finger alphabet.
- *Substituting the Problematic Word*
  If there is not any equivalent sign for a specific word, it should be replaced with a substitute. For example, 'sorun' and 'problem' (both mean 'problem' in English) are synonyms in Turkish in the general sense but in TSL they have only one equivalent sign.
- *Paraphrasing*
  'To paraphrase' means to make an explanation, to restate the sentence in order to give the same meaning in a simpler way. For instance, Turkish word 'hibe' as an equivalence of 'grant' in English can be paraphrased as 'money + help' in TSL.

Figure 3. Paraphrasing the Word 'Hibe' (Grant): Para (Money) + Yardım (Help)

Para (Money) **Yardım (Help)**

Turkish word 'ötanazi' meaning 'euthanasia' in English does not have any equivalence in TSL. Hence, the interpreter may propose finger spelling strategy which includes using signs that correspond to the letters of the word. However, the hearing-impaired person will probably not understand its meaning, even if the interpreter fingerspells it, since there is not any signifier or signified for that word in TSL. Therefore, the most reasonable way to render the meaning of 'ötanazi' is paraphrasing it by using these three words 'legal+ death+ permission' in TSL.

Figure 4. Paraphrasing the Word 'Ötenazi' (Euthanasia): Yasal (Legal) + Ölüm (Death) + İzin (Permission)

Yasal (Legal)

Ölüm (Death)

İzin 1 (Permission 1)

İzin 2 (Permission 2)

4.1.7. Temporary Signs

When a spoken word doesn't have any signed equivalence, a temporary sign can be used for that specific context. As signs are produced by the hearing-impaired people naturally, such solutions should be used only in those particular contexts.

4.1.8. Generalisation

Generalisation can be explained as rendering a specific word by using its superordinate. For instance, if the sign language does not have any equivalence for 'bluefish', the interpreter has to interpret bluefish as 'fish' which is a general term (generic name).

'Finger alphabet' and 'paraphrasing' functioning as a bridge between two languages, are the most common strategies used by sign language interpreters.

However, interpreters' resorting to finger alphabet in long words may lead to loss of time. By increasing the number of interpreting training programmes in Turkey as well as improving the quality of these programmes, interpreters will acquire the language skill to extract the meaning and be able to find the exact equivalence in the target language much more easily. According to Seleskovitch and Lederer (1984, p. 34), "Rendering the meaning of an utterance into another language is actually restating," and in the light of this statement it might be concluded that paraphrasing is also one of the effective ways to render the message in sign language interpreting.

4.2. Simultaneous Interpreting from Turkish Sign Language into Turkish

There are six steps of simultaneous interpreting from TSL into Turkish:

a) *Observing*
At this stage, an interpreter has to see signs, mimics, the position of the body, and the lips as a whole.

b) *Analysing the Meaning of the Message*

c) *Comprehending the Process*
It is related to understanding the content of the message.

d) *Memory and Organisation Processes*
This step includes memorizing the message and organising the manner to render the meaning in the target language.

e) *Completing the Process*
The interpreter completes the process by interpreting the meaning that has been extracted from TSL into Turkish. At this stage, the medium of communication is changed from visual spatial to auditory.

f) *Evaluating the Interpreting*
Interpreters should evaluate the quality of their own interpretations in order to assess whether or not their interpretations can render the meaning and the effect of the source message.

In the light of these explanations, sign language interpreter is expected to use the mental and linguistic skills at the highest level simultaneously. From the process of comprehension to the production process, sign language interpreters exert more effort than interpreters of spoken languages due to the structure and features of sign languages.

5. Concluding Remarks

Strategies used in the sign language interpreting aim at transferring the meaning by taking the target audience into consideration. The improvement of the training on sign language interpreting in Turkey will contribute the interpreters to gain the competence in understanding and reproducing processes of interpreting. To be able to realize this, universities should offer undergraduate and postgraduate programmes on sign language interpreting within the framework of a transdisciplinary approach, in which on-the-job training programmes should also be promoted. By means of sign language interpreting, it is possible to overcome the barriers of communication between hearing-impaired people and the hearing ones.

References

Açan Aydın, Z. (2013). Sign Languages and Aspects of Turkish Sign Language TİD [İşaret Dilleri ve Türk İşaret Dili Özellikleri]. *Hacettepe Üniversitesi Edebiyat Fakültesi Dergisi* [*Hacettepe University Journal of Faculty of Letters*]. 30(1), 73–96. İstanbul: Hacettepe Üniversitesi [Hacettepe University].

Delaporte, Y. (2005). *Dictionnaire de la langue des signes française d'autrefois* [Eski Fransız İşaretlerinin Dili Sözlüğü (Etimolojik ve Tarihsel)]. Paris: Editions du Cths.

Dikyuva, H., Makaroğlu, B. ve Arık, E. (2015). *Türk İşaret Dili Dilbilgisi Kitabı* [*Turkish Sign Language Grammar*]. Ankara: Aile ve Sosyal Politikalar Bakanlığı [Republic of Turkey Ministry of Family and Social Policies].

Donald, M. (1993). Précis of Origins of the Modern Mind, *Behavioral and Brain Sciences*, 16 (4).

Jacobson, R. (1963). *Essais de Linguistique Générale* [*Genel Dilbilim Denemeleri*]. Paris: Minuit.

Lederer, M. (1994). *La Traduction aujourd'hui – le modèle interprétatif* [*Günümüzde Çeviri – Yorumlayıcı Model*]. Paris: Hachette.

Moody, B., Vourc'h A., Giraud, M., Hof, D. vd. (1998). *La Langue des Signes I* [*İşaret Dilleri I*]. Paris: IVT.

Oral, Z. (2015). Türk İşaret Dilinde Karşılığı Olmayan Sözcüklerin Çevirisinde Uygulanan Bazı Stratejiler [Some Strategies Employed in The Translation of Non-equivalent Word In Turkish Sign Language] Edebiyat Fakültesi Dergisi / Journal of Faculty of Letters. Cilt [Volume] 32. Sayı [Number] 2. Aralık [December] 2015.

Pointurier, S. (2014). L'interprétation en Langue des Signes Française contraintes, tactiques, efforts [Fransız İşaret Dilinde Tercüme; Zorlukları, Yöntemleri,

Çözüm Yolları]. Thèse de doctorat en traductologie [Çeviribilim alanında Doktora Tezi]. Ecole Supérieure d'Interprètes et de traducteurs Université Paris 3 Sorbonne Nouvelle.

Ricono, C. (2004). *Regard sur les compétences des interprètes Français/Langue des Signes Françaises* [Fransız Sözlü Çevirmenlerinin Yetkinlikleri Üzerine / Fransız İşaret Dili]. Mémoire de DESS Interprétariat LSF/français, Université Lille 3 [Sözlü Çeviri Alanında Yüksek Lisans Tezi. Lille 3 Üniversitesi].

Sandler, W. and Lillo-Martin, D. (2006). Sign Language and Linguistic Universals. [İşaret Dili ve Dil Bilim]. Cambridge University Press. Aktaran Makaroğlu, B. [In Makaroğlu, B.] (2012). Türk İşaret Dili'nde soru: Kaş hareketlerinin dilsel çözümlemesi [Question in Turkish Sign Language: Linguistic Analysis of Eyebrow Movements]. Yüksek Lisans Tezi [Master's Thesis]. Ankara Üniversitesi [Ankara University].

Seleskovitch, D. ve Lederer, M. (1984). *Interpréter pour traduire* [Yoruma Dayalı Çeviri]. Paris: Didier.

Seleskovitch, D. ve Lederer, M. (2002). *Pédagogie raisonnée de l'interprétation* [(Açıklayıcı) Sözlü Çeviri Eğitimi]. Paris: Didier.

Stokoe, W.C. (1960). *Sign Language Structure: An Outline of the Visual Communication Systems of the Amerikan Deaf* [*İşaret Dilinin Yapısı: Amerikan İşaret Dili Görsel İletişim Sistemlerinin Ana Hatları*]. Vol. 8. Silver Spring, MD: Linstok Press.

Stokoe, Casterline, Croneberg. (1965). A Dictionary of American Sign Language on Linguistic Principles.

Uygur, N. (2013). *Dilin Gücü Denemeler* [Essays on Language Power]. İstanbul: YKY.

Electronic Source

Cuxac, C. (1993) *Iconicité des Langues des Signes. In: Faits de langues,* [*Dillerin ve İşaretlerin (İkonlaşması).* Dil Olguları] n°1, Mars 1993. Motivation et iconicité. (Motivasyon ve İkonlaşma) pp. 47–56. DOI: 10.3406/flang.1993.1034 www.persee.fr/doc/flang_1244-5460_1993_num_1_1_1034

Burak Özsöz M.A.

Marmara University
Department of English Language and Literature

Exploring Linguistic Phenomena in Technical Translation Discourse

Abstract: When describing what technical translation is and what is not, the examination of the theoretical frameworks within which technical translation operates is of the utmost importance. Technical translators with a relevant degree are assumed to have been equipped with the theoretical and conceptual tools which are thought to enable them to effectively deal with previously unencountered situations as they emerge in professional workplace. Yet, with various intended learning outcomes and objectives in the curriculum, it is hardly any doubt that the academic training programs are naturally designed only to familiarize the future translators with the texts that are typically translated in scientific and technical domains.

Likewise, given the enormous diversity of scientific and technical expressions today, scientific and technical terminology is too vast and complex to be fully mastered. Technical discourse being unique, the kinds of competences that may be expected of technical translators are not anything like those to be expected from translators of literary texts. It is not wrong to assume that general translation strategies may not embrace the reality of translation as a form of technical communication, which suggests that without the background knowledge of technical communication being incorporated into the theory and practice of translation in general, technical translators will require a variety of other tools to understand the processes necessary to create a target text involving technical information.

This study aims at revealing how linguistic phenomena which have fed into the developments in translation theory can bring new insight to the technical translator's task. Technical translators produce texts and crafting a new text requires the ability to observe the norms of textuality and to be sensitive to the genre conventions of the target culture. Thus, in dealing with terminology in different kinds of specialized texts, technical translators will need the tools and concepts that come from linguistics and other related (sub) fields, such as textlinguistics, pragmatics, and semiotics.

1. Introductory Remarks

Whether they have undergone an intensive formal training, it might be assumed that there are hardly any barriers for those seeking to enter the translation profession. Today, the increased number of formal institutions offering opportunities in translation and interpreting training can be said to have given rise to a plethora of translators, whether certified or not. Having satisfactorily completed the academic training programs, the translators are considered to have mostly acquired the knowledge and the skills required for proper and acceptable translation. These skills and abilities can be further improved even after graduation, for academic programs equip the would-be translators with the theoretical and conceptual tools which are thought to enable them to effectively deal with previously unencountered situations as they emerge in professional workplace.

As for the scientific and technical translation in particular, it is hardly of any doubt that academic training programs do no more than familiarizing the would-be translators with texts that are typically translated in scientific and technical domains. It can be argued that most professional translators lack a formal scientific and technical training and given the enormous diversity of scientific and technical expressions today, their terminology is too vast and complex to be fully mastered. Scientific and technical discourse being unique, the kinds of competences that may be expected of scientific or technical translators are not anything like those to be expected from translators of literary texts. Likewise, theoretical investigation into scientific and technical translation can be said to have been largely ignored, compared to that of the literature on translation theory in general. It is suggested that theories of translation not embracing the reality of translation as a form of technical communication, there is an urgent need to incorporate knowledge of technical communication into the theory and practice of technical translation (Byrne, 2006; Olohan, 2016). What is suggested in this study is technical translators be exposed to a fundamental linguistic input in order that they can better understand the processes needed to create a text in the target language involving technical information. To this end, the discourse of linguistic phenomena needs to be explored in the discourse of scientific and technical translation. The following part is intended to present information on what is meant by discourse and discourse of scientific and technical translation.

2. The Concept of Discourse

'Discourse' appears to be an umbrella term used for a variety of purposes in various fields of scientific study, including text linguistics, pragmatics, and translation studies. Kocaman (2003, p. 5) maintains that the term expresses a collection of meanings, such as one's point of view; ideology; verbal or written structure and content; idiolect; one's argument or assertion. Although it serves as a more comprehensive term, the notion of discourse is mostly referred to as 'language in use in social interaction' (Van Dijk, 1997, p. 2) and is mainly associated with orality. Doğan (2014, p. 57), however, asserts that discourses, just like language, are manifested in texts and that 'discourse' is not confined to oral production only. Having much to share on the conceptual basis, 'discourse' and 'text' are often used as synonymous terms. Bengi-Öner (2003, pp. 56–59) argues that the concept of discourse is of great prominence in the context of translation process. She further argues that when speaking of what translation is and how translation studies has developed into what it is now, it is in fact the discourse of translation which is being referred to and suggests that those engaged in translation critics do nothing but criticize the translation discourse. She also asserts that to answer the questions in a proper manner regarding what translation studies is, particular attention has to be paid in the first place to the theoretical examination of the discipline. In this sense, this study seeks to highlight the discourse of scientific and technical translation process, with a particular emphasis placed on the discourse of linguistic phenomena, which is essential for translators to have an absolute grasp of knowledge of technical communication in the theory and practice of scientific and technical translation.

3. Discourse and Scientific and Technical Translation

It is aimed herein to provide a general overview of the field of scientific and technical translation and an understanding of what scientific and technical translation discourse is. It is worth noting that all the theoretical concepts and research once suggested and recognized, or now being discussed and developed or thoroughly discredited due to new paradigms or shifting viewpoints, as well as the terms that have yet been coined, are what constitute a sound academic basis for scientific and technical translation, and what may be called the *discourse of scientific and technical translation*. Due to the concerns over communicative purposes, 'science' is paired, as in common usage, with 'technology' just as they are often paired together in translation discourse. To emphasize the standings of science and technology in society and why it might be convenient to group them together, Olohan (2016, p. 16) maintains,

> A key point to be made here, [...] is that traditional views of science as a neutral, objective, value-free activity carried out by disinterested scientists is heavily challenged by much of that scholarship. It is argued that science is a culturally and socially contingent activity, which can be ideologically and politically driven. Likewise, technology, often portrayed as the means by which many of the world's problems and our human imperfections can be fixed, must also be examined closely for its motivations, influences and effects. [...] We can adopt the convenience of grouping science and technology together, while recognizing that the nature of the domains and the dynamic relationships between them can be understood in different ways.[1]

1 In what follows, as was suggested by Olohan, the two terms are grouped together as 'technical translation'.

There is also the question of the nature of the relationship between specialised and technical translation, for a text having specialised terminology is wrongly assumed to be technical. To emphasize the distinction, Byrne (2006, p. 3) argues,

> [...] Just because there is a specialised terminology, doesn't make something technical. In discussing technical translation, it is useful to make the distinction between specialised and technical translation. For example, religion has a very specific terminology and very definite conventions, styles and document structures but it is never regarded as 'technical'.

Technical translation is, on the other hand, viewed by Newmark (1988, p. 151) 'as one part of specialised translation'. Erten (1997, p. 18) states that to a great extent, the problems encountered in translating technical and scientific texts are no different than those encountered in performing a translation task in any field, whether specialised or not. She still does not refrain herself from asserting that technical texts are content-wise that are mainly less familiar to many a translator, leading to the translators' lack of interest in the translation of such texts. It is not intended here to shift the focus to the general status of technical translation in Translation Studies, but still there is a need to emphasize the distinctive features that differentiate technical translation from the other forms of translation, including the terminology, which deserves a special treatment. In translating non-literary texts, for which 'technical translation' is often used as a cover term, Rogers (2015, p. 4) states, "It might be assumed that the scope of any study of 'non-literary' translation on the basis of terminology is highly constrained, being limited to two aspects: (a) in areas of discovery and invention, morphological issues of (primary) term formation, and of (secondary) term formation in a relatively culture-free zone; and (b) in more established areas, lexical, semantic issues of equivalence." Newmark (1988, p. 151) also views technical translation as potentially non-cultural and argues that technical translation is mainly distinguished from the other forms of translation by terminology.

Still, the increasing information gap between various cultures is yet to be considered. New science and technology invented by more developed nations are mostly transferred to less developed nations through translation. In this sense, concepts and terms created in non-literary, scientific and technical domains can be said to be confined to certain cultures. Although it may be true that terminology accounts for an important aspect of scientific and technical texts, Byrne (2006, p. 3) points out, "Technical translation is not all about terminology." Scientific and technical texts may rely on an array of carefully chosen words, but the problems of terminology are closely linked with the major problems of 'text type' and being aware of the characteristics of different types of texts is considered as one of the first steps of translation competence (Yazıcı, 2007, p. 98). Yet, compared to literary translation, technical translation exhibits a much greater conceptual distance between the translator and the text and it would not do for the technical translators to rely on their general world knowledge only. They need to acquire a good amount of subject-matter knowledge if they are to produce translations of high quality (Byrne 2006, pp. 5–6).

While, for instance, translating texts with the informative function, technical translators should be well informed about the characteristics of scientific and technical texts that influence the text comprehension, which is associated with the *genre* of the instructional texts. Even though some texts are produced to inform the target readers of the new technology that is involved, translators should know that technical translation is a form of technical communication, which fulfills a range of communicative purposes. Examining the genre of technical instructions, Olohan (2016, p. 57) discusses that a key criterion for designers of any technical documentation is *usability*. Instructions enabling users to install and operate products should not be too lengthy to inhibit the operative function to be fulfilled. To emphasize that technical translation is more than simply transmitting information, Byrne (2006, p. 10) argues,

> The challenge for technical communicators is to ensure that all of the relevant information is indeed conveyed but also that it is conveyed in such a way that the readers can use the information easily, properly and effectively.

In translating literary texts, there is the evidence of the direct involvement of three parties: the author, the translator, and the reader. However, given that a large portion of the non-literary texts are not produced specifically as a result of the author's desire to be translated into another language, Nord (1997, p. 21) claims that the source text author would not be 'an immediate agent in the translational action', and therefore, would not have any immediate involvement in the translation process, emphasizing the fact that the arguments over achieving precision or accuracy in creating a target text either on the terminological or syntactic level can be resolved with a new understanding of the centuries long debate on source oriented versus target oriented translation. Even though the proponents of both camps will not be here revisited in detail, it might help to mention a few recent approaches of both sides from the specific perspective of technical translation theory so as to highlight the relation between a source text and a target text. Byrne (2006), for instance, rather than inquiring whether it is the source text being the sole guiding principle for the production of a target text, can be said to have provided a different insight into the discussion, suggesting that technical translation should be approached from the perspective of the *usability* of the document that is created. According to Olohan (2013, p. 430), Byrne has more of an inclination to focus on the production of translations that correspond to readers' cognitive capabilities, without requiring target readers to spend much effort to fulfill their communicative functions. It can, therefore, be concluded that technical translators' top priority has shifted from 'precision' to 'comprehensibility' as an essential part of a decision-making process. Hann (1992, p. 7), for instance, seeks to define this process and suggests that it involves 'the selection of the correct target-language rendering from a number of different, context dependent alternatives.' While, in traditional approaches, the source text is viewed as the starting point of the entire translation process; a translation is, as suggested by Toury (1995, p. 26), a fact of the target language that hosts it, meaning that translation of technical texts can only be judged in the context of the norms of the target text to be created. The following part is intended to highlight how translation discourse incorporates some of the components of linguistic competence when dealing with technical texts.

4. Linguistic Discourse and Technical Translation

It has now been strongly established that linguistics and translation studies do not represent a hierarchical relationship. Translation studies being interdisciplinary in nature, linguistic theory, and insights from other disciplines have fed into developments in translation theory. What is sought here are some areas in technical translation where linguistic discourse can bring insights to the technical translator's task. To this end, it might be useful to refer in advance to a few reasons why early developments in linguistics were relatively of little interest to translation theorists.

Mostly in the first half of the previous century, when the linguistic description confined itself to a single language system and morpho-syntactic studies were of the main constituent of major analysis to the exclusion of semantic investigation, translators and linguists were not speaking about the same thing since interlingual translation involves two languages and two cultures, which are obviously different from each other.

In the second half of the century, efforts that rely on more than a single language system were made to develop a general linguistic theory of translation. Catford (1965, p. 1), for instance, suggested, "Translation is an operation performed on languages: a process of substituting a text in one language for a text in another."

Although Catford considers languages in their natural social settings in which they operate and her work is a good representative of the attempts to describe 'translation' systematically from a linguistic point of view rather than merely applying the findings of linguistics to the elements within the text to be translated; much of the discussion draws on the structural contrasts between language systems and the model largely ignores the cross-cultural barriers in communication despite the fact that communication takes place in social situations. Besides, the model fails to go beyond the sentence level to incorporate the text as a unit of meaning. Hatim and Mason (1990, p. 30) emphasize that in this discussion, the communicative value of the items in particular contexts under investigation is not taken into consideration. Despite Catford's work and the like concentrated on the word and sentence levels, the linguistic influence on translation remained substantial since the translation theorists adopted a variety of concepts from the other levels of linguistic analysis. Among such analyses are those of 'discourse and pragmatic' which came to prominence to translation studies in 1990's and they still have much to offer to explain translation as a process or a product. It is now apparent that isolated sentences can no longer be treated as the basic unit of linguistic investigation and that language cannot be studied in isolation from the context where the communicative intention of the language user is truely revealed. The translator theorists influenced by the developments in discourse analysis have shifted their focus of attention from the smaller units of language (lexical or syntactical equivalences) to a perspective which considers the text to be translated as a whole within its communicative situation.

No text can be studied apart from its context and 'context', most of whose discussion is from outside Translation Studies, can be defined as various aspects of extra-linguistic reality, which plays an essential role in the construction of meaning. In translating a text, including technical texts, a translator should consider the translation process as 'a whole set of extra-linguistic criteria' which support the linguistic elements. Bell (1991, p. 110) posits that *text comprehension* consists of reconstructing the context from the words of the source text and emphasizes the importance of setting aside all the words of the original text to see the picture clearly. To him, what has to be communicated is the idea or the picture, not equivalents of the actual words. Melby and Foster (2010, p. 4) state,

> When assigning meaning to a source text, one must not limit one's focus only to those aspects of the situation that are linguistic in nature. Many nonlinguistic variables can help determine the meaning of source materials, including who wrote the material, what situation that person was in, and surrounding cultural events that may clarify the intentions of the writer.

From the aforementioned discussion on the context analysis made for the translation purpose, it can be restated that linguistic and extra-linguistic criteria should be employed together. *Culture* is such an extra-linguistic feature of a text and cross-cultural communication can sometimes be problematic in technical translation. Technical translators' task can be cited as an example when they have to deal with instructional materials – print or non-print – used for instructing users. While a substantial part of instructions has a verbal component, designing wordless instructions with images can be an alternative way. Cross cultural understanding of such images may be problematic in that images are merely used to represent concepts. Concepts represented by images embody symbolic and socio-cultural conventions. Olohan (2016, p. 50) emphasizes,

> The use of diagrams and other forms of images may be carefully considered by technical writers and document designers at the time of authoring, but on many occasions, it is assumed by clients and others that translated texts can simply reuse the same images. Sometimes this is unproblematic, but translators and clients should also be aware of the potential for cultural specificity and of the necessity to make judgements about cultural appropriateness, extending also to use of specific colours [...].

Professional translators should not limit their contextual focus to the text surrounding a word or phrase; nevertheless, it seems necessary for technical translators to understand in the first place that words are linguistic signs and interlingual translation is only possible through the interpretation of a set of linguistic signs of one language by means of some other linguistic signs of another language. It is only through the study of the knowledge of linguistic signs that a technical translator is able to understand the aspects of context that go beyond text. Melby and Foster (2010, p. 7), for instance, maintain that not all the things relevant to translation can be reduced to text and that there is often more than the task of translating, which they call 'paralinguistic information', such as cultural awareness and technical knowledge. This is why technical writers, technical professionals, or technical translators should be well aware of the fact that language is a linguistic coding which is made up of a set of linguistic signs supported by cultural specific nonlinguistic signs and that to ensure the usability of information, the textual content of technical documents is often reinforced by graphics and images, which are regarded as extra-linguistic signs. Thus, the random substitution of one image or symbol with another in the target language will not have the same impact on the target reader.

A sign, be it linguistic or extra-linguistic, is only meaningful when it comes to interact with other signs in a system, in other words, when a new sign is incorporated into a system, it will influence the communicative value of the rest of the signs. A sign, regardless of whether linguistic or nonlingistic, is not to be isolated from the system to which it belongs. To prevent misunderstanding, it is not meant to state that a sign which stands all by itself, outside a system does not hold any meaning. However, a sign within a context helps to avoid semantic confusion (Özsöz and Canseven, 2013, p. 759).

From the semiotic point of view, translation is not language-based but sign-based, dealing with the interpretation of sign systems both verbal and non-verbal. To support this view, Gorlée (1994, p. 13) states,

> Point of departure of semiotic linguistics is the Saussure-based claim that all language is a system, a coherent semiotic structure, and that consequently, all text can be described and analysed semiotically.

Unlike Saussure's two-part mental linguistic unit (signifier and signified), Pierce dwells upon 'sign' as being made up of three elements, namely, representamen, object, and interpretant. He defines 'sign' as a conception that exceeds natural languages. In an attempt to develop a semiotic model for translation theory, this triadic conception of sign was used[2], for Pierce was not primarily interested in how language worked and how it was used by people in communication, but concerned himself with understanding how signs and sign-action work in the entire universe to suggest that translating is not existentially defined by language (Stecconi, 2007, p. 18). With this three-part model of sign, it is easier to see that translating can be cited as something that "we do *to* words" as well as to other signs, not something that "we do *with* words". Thus, technical translators, when confronted with translating that involves nonverbal signs, such as colours and still images on the screen, need to look beyond words to make real sense of what they do. Due to the fact that translation theories are mostly based on verbal language, it can be argued that a general theory of translation that embraces verbal and nonverbal signs altogether is called for.

5. Concluding Remarks

If technology is assumed to be the *worldly* face of science that encompasses humans' intent to manipulate nature to solve their problems in life, it may then be concluded that technology embraces semiotically and linguistically all discrete units that exist in the entire universe of human consciousness, resulting in an inconceivably large number of discourse representations used by natural languages to relate themselves to a universe of such quality. Considering the composite text types and discourses technology encompasses and the technical terms which are only meaningful within a specific industry, and given that technical documents are often produced by technical writers heavily depending on the consistent use of a technical terminology with all of which technical translators cannot be expected to be familiar, it is obvious that technical translators' task is not an easy one.

Translators make decisions mostly based on their training and experience, but it is clear that they need to consult experts to understand source language terms they are unfamiliar with. Furthermore, they will have to fill the terminological gaps in the target language. Nonetheless, technical translators' task is not to be reduced to

2 For further information, see Van Kesteren (1978).

an understanding as if it is simply an issue of knowing the terminology and coining of new terms to label concepts. They produce texts and crafting a new text requires the ability to observe the norms of textuality and to be sensitive to the genre conventions of the target culture.

Translators with an approved degree or certificate are assumed to have been mostly equipped with the theoretical and conceptual tools that help them deal with heterogeneous range of text types and discourses. Among such tools and concepts are those that come from linguistics and other related fields, such as textlinguistics, pragmatics, and semiotics. Training future translators means introducing them to what is central to translation task, as well as raising their linguistic awareness, both in the source and target language. In translation studies, translation of texts of specific domains is where linguistic analysis is mostly benefited from, since such texts, due to their informative function, allow for linguistic generalizations (Bulut, 2002, p. 20).

References

Bell, R. T. (1991). *Translation and Translating: Theory and Practice.* London & New York: Longman Group.

Bengi-Öner, I. (2003). Söylem ve Çeviri [Discourse and Translation]. In A. Kocaman (Ed.), *Söylem Üzerine* [Upon Discourse] (pp. 56–81). Ankara: Metu Press.

Bulut, A. (2002). *Çeviride Dil ve Metin* [Language and Text in Translation]. Istanbul: Istanbul University Press.

Byrne, J. (2006). *Technical Ttranslation: Usability Strategies for Translating Technical Documentation.* Dordrecht: Springer.

Catford, J. C. (1965). *A Linguistic Theory of Translation: An Essay in Applied Linguistics.* London: Oxford University Press.

Doğan, A. (2014). *Sözlü ve Yazılı Çeviri Odaklı Söylem Çözümlemesi* [Discourse Analysis for Interpreting and Translation Research]. Ankara: Siyasal.

Erten, A. (1997). Approaches to Technical Translation. *Çeviribilim ve Uygulamaları.* 7, (pp. 17–23).

Gorlée, D. (1994). *Semiotics and the Problem of Translation.* Amsterdam and Atlanta: Rodopi.

Hann, M. (1992). *The Key to Technical Translation.* Amsterdam and Philadelphia: John Benjamins.

Hatim, B. and Mason, I. (1990). *Discourse and the Translator.* London: Longman Group.

Kocaman, A. (2003). Dilbilim Söylemi [Linguistic Discourse]. In A. Kocaman (Ed.), *Söylem Üzerine* (pp. 1–11). Ankara: Metu Press.

Newmark, P. (1988). *A Textbook of Translation*. New York: Prentice Hall.

Nord, C. (1997). *Translating as a Purposeful Activity: Functionalist Approaches Explained*. Manchester: St. Jerome.

Olohan, M. (2013). Scientific and Technical Translation. In C. M. Valera and F. Bartrina (Eds.), *The Routledge Handbook of Translation Studies* (pp. 425–37). London and New York: Routledge.

Olohan, M. (2016). *Scientific and Technical Translation*. London and New York: Routledge.

Özsöz, B. and Canseven, C. (2013). Dil Göstergesinin Devingenliği [Dynamism of Linguistic Sign]. In *Book of Proceedings of 13. International Language, Literature and Stylistics Symposium: Simple Style* (pp. 755–762). Kars, Turkey: Kafkas University.

Rogers, M. (2015). *Specialised Translation: Shedding The 'Non-Literary' Tag*. UK: Palgrave Macmillan.

Stecconi, U. (2007). Five Reasons Why Semiotics Is Good for Translation Studies. In Y. Gambier, M. Shlesinger and R. Stolze (Eds), *Doubts and Directions in Translation Studies* (pp. 15–26). Amsterdam and Philadephia: John Benjamins.

Toury, G. (1995). *Descriptive Translation Studies Beyond*. Amsterdam and Philadelphia: John Benjamins.

Van Dijk, T. A. (1997). The Study of Discourse. in T. A. Van Dijk (Ed.), *Discourse as Structure and Process. Discourse Studies: A Multidisciplinary Introduction*. Vol. I. (pp. 1–34). London: Sage.

Van Kesteren, A. (1978). Equivalence Relationships between Source Text and Target Text: Towards a Typology on the Basis of Semiotics. In J. S. Holmes, J. Lambert, R. Van den Broeck, M. Janssens and A. Lefevere (Eds.), *Literature and Translation: New Perspectives in Literary Studies* (pp. 48–68). Leuven: Acco.

Yazıcı, M. (2007). *Yazılı Çeviri Edinci* [Translation Competence]. İstanbul: Multilingual.

Electronic Source

Melby, A. K. and Foster, C. (2010). Context in Translation: Definition, Access and Teamwork. *The International Journal for Translation & Interpreting Research*, 2(2) retrieved from http://trans-int.org/index.php/transint/article/view/87/70 on 21.6.2016.

Gözde Begüm Uyanık M.A.

Marmara University
Department of Translation and Interpreting

Translation and Interpreting in Sports Contexts

Abstract: This study deals with a relatively recent field in translation: Sports Translation, and handles it as a field of specialization. The study first divides Sports Translation into fields by introducing them in terms of its types (*mode*) and contexts (*modality*). Secondly, it handles as well as discusses Sports Translation as a field of specialization by defining the role of the Sports Translator, and by observing its technicality in terms of its use of technical terms and subject matter. Moreover, the chapter also tries to conceptualize Sports Translation as a type of Translation. Finally, the last part of the study suggests strategies for translation of sports language by mentioning two strategies related to the terminology – namely *borrowing* and *calque* – as well as strategies regarding the text type with examples provided from sports news and rules book.

1. Introductory Remarks

Sports Translation is the term holding a wide range of definitions. Before touching upon those various definitions, it is first and foremost necessary to state that by using the term 'Translation/Translator' in this study, both translation/translator and interpreting/interpreter are meant[1]. Translation used herein with the capital 'T' is the umbrella term for both written and oral renderings in Sports Translation.

The definitions of Sports Translation could be regarded as:

a) every type of Translation within the sports field,
b) Translation service given to the people within the sports industry, and
c) a sub-type of community interpreting.

These definitions may vary according to the *mode* and *modality.*

2. The Key Concepts: Mode and Modality

'Mode' and 'modality' determine the types of Sports Translation. The former stands for *the way Sports Translation is performed,* while the latter stands for *the contexts where Sports Translation is performed.*

---

1 The capital 'T' is used in the same manner as Pöchhacker (2004, p. 9) does for 'Translation' so as to mean both Translation and Interpreting.

2.1. Mode

Modes of translation in the field of sports are, first of all, bidirectional; that is, written and oral renderings in terms of the way they are performed – as explained in Table 1.

Table 1. Modes in Sports Translation

Modes in Sports Translation	
Written Sports Translation – press releases – rules books – websites of sports teams/ international sporting events – sports news …	**Oral Sports Translation (Interpreting)** * *simultaneous* – draws in international sporting events * *consecutive* – press conferences – in-field interpreting …

In sports field, written renderings could be press releases, rules books, translation of the websites of sports teams or international sporting events, translation of sports news, and so on. As for the oral renderings in sports, simultaneous as well as consecutive modes are performed. Simultaneous mode could be used in draws of international sporting events (World Cup Draw, for instance), whereas consecutive mode could be observed in press conferences or in-field interpreting.

2.2. Modality

Modality involves the contexts. Contextual differences are closely interrelated with the types in translation (such as medical translation, legal translation, and so on) as well as interpreting (such as medical interpreting, court interpreting, community interpreting, and so on) in general. The same is also true for Sports Translation as a sub-type of Translation.

Sports Translation could be divided into two in terms of the way the Translation services are provided. Firstly, they could be about sports; in other words, the context of the Translation services could be *about* sports. Secondly, these services could be provided *for* the people within the sports industry such as sportsmen, trainers, sports teams, and so forth.

To sum up, while Sports Translation could be regarded according to the way it is performed (mode), it could also be regarded according to the contexts where it is performed (modality).

3. Sports Translation as a Specific Field

There are two important models in communication; namely *transfer (conduit)* model and the *social-interactionalist* model. While the former is monological, the latter model is dialogical. In the social-interactionalist model, "the speaker interacts with his/her physical and social environments, with interlocutors and [*sic*] situational and cultural contexts; messages are co-produced in dialogue" (Linell, 1997, p. 52). Therefore, an interpreter is the person who provides communication within these situations and contexts and between people or groups belonging to different cultures.

Being both a language expert and a cultural expert, the most important task of a Translator is to provide communication as accurately and smoothly as possible between parties with different cultural backgrounds[2]. However, a Sports Translator is also expected to have thorough knowledge of the terminology and the subject matter of the technical field where Translation task is performed.

The reason why the term 'technical' has been used in Sports Translation is intentional, for sports can be regarded as a technical field with its own terminology, and the field of sports could cover a variety of fields thanks to its interdisciplinary nature as well. For example, for the last two decades 'sports law' has emerged as a sub-discipline of law all over the world and in Turkey. Sports law deals with cases and suits of the wide range of clients demanding representation and legal advice. This field requires knowledge of several sub-disciplines of law. The distinctive feature of the sports law comparing to other sub-disciplines of law is that the clients with whom sports lawyers deal are mostly professional or amateur players, coaches, referees/officials, and governing bodies of the sports industry. Even more broadly, sports lawyers' "[...] representation might extend to sports broadcasters; sports equipment manufacturers; sports medicine care providers; businesses that sponsor athletic events or athletes; and concessionaires who serve food and drink to fans at games."[3] Thus, sports lawyers may find themselves dealing with a wide range of legal suits from medical malpractice suits to intellectual property law suits and arbitration. As this is the case, it could be rightful to say that a sports lawyer works in every sub-discipline of law dealing with wide range of clients. No matter how skilful a sports lawyer is, he/she may not have a good command of a foreign language. This is when a Translator is required. A sports Translator could be the one who translates legal documents for the sports lawyer or the client, or he/she could be the one who provides assistance to an attorney and his/her client to communicate with one another. The profile of the sports-related people within the sports industry certainly varies, so do their language combinations. Whenever a person within the sports industry requiring legal assistance consults to a sports lawyer and whenever they could not agree on a language to communicate one another, they need an interpreter. The topic for this type of interpreting is mainly related to sports; thus, sports interpreters step in this subject matter. However, since

2 For further information, see Kondo and Tebble (1997, p. 153).

3 This information was retrieved from the website of American Bar on 24.02.2016.

it is also related to the field of law, sports interpreters should have knowledge about legal terms in addition to the subject matter.

The link between sports and medicine is also worth mentioning to reflect the interdisciplinary nature of sports. Even a person outside of Translation Studies or a person who has not been involved with Translation at all would be aware of the high possibility of players' getting injured during games and trainings more than an average person does since this is the nature of their profession. Players staying in a foreign country as members of a local team or for an official sporting event (the world championship, for instance) may get injured and need medical assistance. Provided that they do not know that country's language, they would need an interpreter to get medical assistance. Again, a sports interpreter should be there to provide the communication between the players and healthcare professionals. In this case, sports interpreter must have thorough knowledge about medical terminology and its subject matter.

In both cases, the need for knowledge of field-specific terminology (such as that of medicine, law, and so on) is obvious. In Uyanık's study (2015) on sports interpreting, 30 sports interpreters were requested to answer a questionnaire with open-ended questions on the difficulties they faced during interpreting. The results of this study are presented in Graph 1 (Ibid, p. 82).

Graph 1. The Difficulties Faced During Sports Interpreting

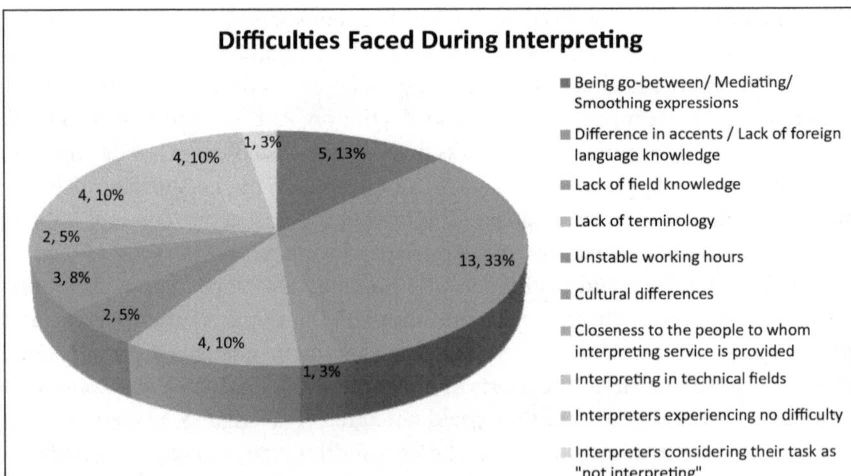

The most prominent two findings of this study are 'the difference in accents / lack of foreign language knowledge' (33%) and 'being go-between / mediating /

smoothing expressions' (13%). Apart from these two findings, 10% of the sports interpreters participated in the questionnaire indicated the discomfort of 'lack of terminology' during their interpreting performances; likewise, 10% of them stated that they suffered from performing technical consecutive interpreting, that is 'interpreting in technical fields'[4] (Ibid.). As this study suggests, the knowledge of technical terminology, namely those of sports and the related fields, is essential for oral renderings in sports as well as those of written.

Sports interpreting also bears significant similarities to community interpreting. One can easily claim and is rightful to say that overcoming aforementioned difficulties is in fact community interpreters' task, for 'community interpreting' is defined in Routledge Encyclopaedia of Translation Studies (Wadensjö, 2001, p. 33) as

> [...] the type of interpreting which takes place in the public service sphere to facilitate communication between officials and lay people: at police departments, immigration departments, social welfare centres, medical and mental health offices, schools, and similar institutions.

What a sports interpreter actually does in legal and healthcare issues is, therefore, community interpreting, which is true in many respects. In fact, sports interpreting could be assessed and handled as a sub-field of community interpreting. At which point it differs from community interpreting is related to for whom this community interpreting service is performed[5].

As is explained, the necessity of medical Translation as well as legal Translation in the field of sports is foreseeable at the first glance. However, Sports Translators may also find themselves in unexpected situations. An observation during a sporting event may provide an insight to the subject: The 'FIBA[6] U18 European Championship Men' was held in Konya, Turkey in 2014[7]. Before the games in the tournament began, each team had two days for training. During this period, the national team of England faced a problem relating to the customs clearance. They had brought jerseys on which the numbers and names were printed incorrectly,

---

4 Herein 'technical field' refers to the 'specific field'.

5 For further information, see Lidi Wang and Jie Zhang (2011, pp. 263–279), 'Community Interpreting in China Since the Beijing Olympics 2008 – Moving Towards a New Olympic Discipline?' for the development and changing the face of community interpreting in China after Beijing Olympics (2008) as an indication of the effect of sports interpreting on community interpreting.

6 For further information, see the website of Fédération Internationale de Basketball; The International Basketball Federation.

7 For further information, see the official website of FIBA U18 European Championship Men on 23.2.2016.

and if they had played with those jerseys during the tournament, they would have been given a penalty fine by FIBA for each game they would play. Being unaware of the customs procedures followed in Turkey, they had ordered new jerseys from China to Turkey to be sent to the general manager of the team. Thus, the jerseys were detained by the customs, and the team attaché[8] of the England U18 team had to talk with the customs officials on the phone. However, the team attaché was not prepared for such an unexpected situation; namely, knowing the customs and trading terms. After communication was established, it was understood that the earliest time England U18 team could get the jerseys would be after the tournament, for most of the games would be played during Ramadan Holiday time. To prevent England U18 team from getting a penalty, something had to be done, and the team attaché found herself in a situation where she looked for a local craftsman to do the patching on the jerseys. She contacted with that person and solved the problem somehow. This was something unforeseeable by the team attaché as well as the local organization committee. In short, as a problem solver, sports interpreters could find themselves in different situations and should be prepared for them. Whether foreseeable or unexpected, Sports Translators have to apply some strategies so as to overcome difficulties faced within Translation process.

4. Strategies for the Translation of Sports Language

The strategies for the translation of sports language can be examined under two categories: the strategies related to the terminology and those regarding the text.

4.1. Strategies Related to the Terminology

Although the number of sports-related terms in Turkish language is quite few, sports is a field of specialization (Bulut, 2011, p. 64); in other words, a specific field. Loan words entered Turkish language by means of translation in many specific fields (such as medicine, law, trade, and so on), and this could be observed within the sports field as well. One can notice two strategies used in translation of sports terminology: *borrowing* and *calque*. Borrowing and calque are the terms used by Jean-Paul Vinay and Jean Darbelnet in 1958[9] as two of the seven procedures for translation.

---

8 Team attachés are the contact people and responsible for a team. They may be language assistants, as well. They maintain communication between parties and what they mostly do is interpreting. Therefore, they are regarded as sports interpreters in this study.

9 For further information, see Vinay, Jean-Paul and Jean Darbelnet's (1958/1995) *Comparative Stylistics of French and English: A Methodology for Translation*.

4.1.1. Borrowing

Vinay and Darbelnet (1958/1995, in Venuti, 2004, p. 129) assert that borrowing is the simplest of all translation methods. This is because terms in the source language are directly transferred into the target language. This translation method is used to fill a semantic gap within the target language. Jeremy Munday (2012, p. 86) states that there is much borrowing of terms in some technical fields (such terms as *computer, internet,* and so on from English to Malay). Similarly, borrowing sports terms could also be observed in Turkish language, by adapting their pronunciation into Turkish. Some examples of the borrowed sports terms in Turkish are as follows:

Football

football: futbol	*goal*: gol	*penalty*: penaltı	*forward*: forvet
back: bek	*offside*: ofsayt	*corner*: korner*	*free kick*: frikik*

Basketball

basketball: basketbol	*period*: periyot	*dribbling*: dripling	*rebound*: ribaunt
assist: assist	*block*: blok	*man-to-man*: mentümen*	*disqualification*: diskalifiye

Volleyball

volleyball: voleybol	*block-out*: blok aut	*defence*: defans	*attack*: atak
passeur: pasör	*antenna*: anten	*service*: servis	*scoreboard*: skorbord

* *These terms have also calque usages.*

Note that these Turkish words are pronounced exactly how they are written.

4.1.2. Calque

Vinay and Darbelnet (1958/1995 in Venuti, 2004, p. 129) describe a calque as "[...] a special kind of borrowing whereby a language borrows an expression form of another, but then translates literally each of its elements." In the light of this definition, there are two kinds of calque: *lexical calque* and *structural calque.* While the former is not intervened with source language structure and respects the structure of the target language, the latter introduces a new structure to the target language; both introducing a new mode for expression. The uses of these two kinds of calque can also be observed in sports language. Some examples of lexical calque in Turkish are as follows:

Football

corner kick: köşe vuruşu	free kick: serbest vuruş	back pass: geri pas

Basketball

jump ball: hava atışı	basket: sepet	bench: takım sırası
backcourt: geri saha	front court: ön saha	free-throw: serbest atış
dead ball: ölü top	live ball: canlı top	shoot clock operator: şut saati görevlisi

Volleyball

double contact: çift vuruş	husband/wife: karı-koca	flare: parlama

Both procedures for translation – namely, borrowing and calque – become integrated in the target language and become a part of that language after a period of time (Ibid). While translating or interpreting, a sports Translator must know how to render sports terms. If he/she does not know that it is a sports term, he/she could render that term literally, thus, resulting in a loss in meaning in the target language. Sports Translator must be aware of this and render the terms accordingly to succeed in transferring the meaning to the parties to the communication.

4.2. Strategies Regarding the Text

As mentioned earlier, the examples of written sports translation are press releases, rules books, websites of sports teams, and/or international sporting events, sports news, and so on. In each case, translation strategies should be different, for each text is addressed to different target audience for different purposes. When a sports Translator is expected to translate/interpret a text, he/she should consider the specific purpose (*skopos*)[10] for the Translation. With respect to his/her specific purpose of translation, a sports translator should translate the given text accordingly. This can be exemplified with the translations of two different text types – sports news and rules book – together with their strategies.

4.2.1. Strategies in Translation of Sports News

In sports news, sports language differs from other types of language use in that it is generally composed of poetic discourse and creativity. According to Tahsin Yücel (2004, p. 17) a football writer is a kind of poet. Moreover, a poetic discourse, especially the use of rhythms, is mostly seen in his/her writing. Yücel exemplifies this as follows:

10 See Hans J. Vermeer's (2004) *Skopos Theory and 'Skopos and Commission in Translational Action.*

Example 1
"Denizli kalesini golle doldurdu, üç puanı zorlanmadan buldu." (Ibid.) **
Paraphrase: Denizli (a Turkish football team) scored so many goals that it easily won the game and got three points.

Example 2
"Paris Saint-Germain golü erken buldu, CimBom'un rüyası 20 dakikada soldu." (Ibid, pp. 17–18) **
Paraphrase: Paris Saint-Germain scored an early goal in the twentieth minute of the game, which faded away Galatasaray's (a Turkish football team) hopes of winning.

*** So as to preserve the rhythmic essence of discourse, the first two examples were not translated into English – only their meanings were given.*

Apart from the poetic discourse, the main objective of sports news is to draw readers' attention through creativity. The respective examples are as follows:

Example 3
"Fenerbahçe'nin içine girip bir türlü çıkamadığı *tencereye* Bursa *cenderesi* deniyordu." (Ibid. p. 18)
Paraphrase: 'Cendere' [mangle] is figuratively used to refer to the hardship Fenerbahçe (a Turkish football team) faced against Bursa (also a Turkish football team), while 'tencere' [pot] symbolizes that Fenerbahçe got stuck against this hardship.

Example 4
"Telekom Kazan'da boğuldu."[11]
Paraphrase: 'Kazan' is used in the example as a pun (a kind of wordplay), for it refers to both 'Unics Kazan Basketball Club' and the 'boiler' (in Turkish), into which Türk Telekom Basketball Club fell and got drowned.

The third and fourth examples involve wordplays in Turkish. Therefore, it is not an easy task for a sports Translator to render the meaning with the same effect in a foreign language other than Turkish. As a result, it can be stated that the uses of wordplays as well as rhythmic language are highly observed in sports news writing, as in other languages' sports news writing. Thus, a sports Translator needs to find a solution so as to create a similar effect in the target language. It can be

11 The headline was retrieved from the web page of Hürriyet (dated 4.2.2009) on 27.02.2016.

concluded that sports translation necessitates creativity in target language, as it does in sports writing discourse in the source language, which is frequently encountered in literary discourse as well.

4.2.2. Strategies in Translation of Rules Book

Sports is a specific field. There are rules for each sports branch so as to standardize and regulate them throughout the world. Therefore, rules books are prepared and they have prescriptive, binding features. Each person before exercising a particular branch of sport – either professionally or unprofessionally – agrees that these rules are binding; therefore, they act accordingly during the game. When a player acts against the general principles and/or the rules of the game, some penalties – as sanctions – are imposed on the player by the officials according to the rules book. Consequently, to some extent, this is similar to the ruling of the law in terms of sanctions which are imposed by the judges. in accordance with the rules.

Two articles are taken as examples from FIBA 'Official Basketball Rules 2014' so as to illustrate the aforementioned similarities, in which the uses of legal language could be observed. In her article on the characteristics of English and Turkish Legal Languages, Ayfer Altay (2002, p. 27) states,

> the helping verb 'shall' does not express the future tense in law discourse although it generally expresses the future tense. In daily language, it is used with the first-person singular and the first-person plural, but in the legal language, it is used to express obligation, order, explanation, and notification.

A similar usage of 'shall' can be observed in the following article in FIBA Official Basketball Rules Book (2014, p. 34): "Any number of fouls may be called against a team. Irrespective of the penalty, each foul shall be charged, entered on the scoresheet against the offender and penalised accordingly." It is clear that this article regarding the fouls in the rules book shows similarities with the legal language usage. The following examples clearly show this similarity[12]:

Example 5

Source Text

"Any number of fouls may be called against a team. Irrespective of the penalty, each foul *shall* be *charged, entered* on the scoresheet against the offender and *penalised* accordingly" (Ibid.).

12 The examples taken from the rules book are original and they have been translated by the writer of this chapter accordingly. The italics in the examples are used for emphasis.

Target Text
"Bir takıma herhangi bir sayıda faul verilebilir. Cezasına bakılmaksızın her faul *değerlendirilir*, sayı cetvelinde faul yapan takımın hanesine *kaydedilir* ve kurallara uygun şekilde *cezalandırılır*."

Example 6
Source Text
"The offensive player, whether on the floor or airborne, *shall not cause* contact with the defensive player in a legal guarding position by:

Using his arms to create more space for himself (pushing off).

Spreading his legs or arms to cause contact during or immediately after a shot for a field goal." (Ibid, p. 35)

Target Text
"Zemindeki veya havadaki hücum oyuncusu;

Kendisi için daha fazla alan yaratmak üzere kollarını kullanarak (iterek),

Sahadan yapılan bir atış sırasında veya atışın hemen akabinde temasa neden olacak şekilde bacaklarını veya kollarını açarak, legal savunma pozisyonundaki savunma oyuncusuyla temasa *neden olamaz*."

As seen in the examples of both source and target languages, it can be stated that when translating a rules book or this kind of another text, a sports translator must pay attention to this kind of legal language use. In other words, he/she should pay attention to the grammatical structures as well as language uses in the source and target languages. The use of languages might show differences in the source and target cultures; however, when translating, a sports translator needs to find correspondences within the target language system. Moreover, he/she should be aware of the source texts and the target texts as well as their audiences.

5. Concluding Remarks

Sports is a specific field in itself so is sports Translation; the prominent characteristics of which can be summarized as:

- There is no single definition for sports Translation; it may vary according to the contexts and for whom it is performed along with the way it is performed (mode and modality).
- Knowledge for field-specific terminology is of utmost importance to sports Translators.
- Sports Translators, especially sports interpreters, should have problem solving skills, for they could easily find themselves in unforeseeable situations.

- Thanks to the interdisciplinary nature of sports, sports Translation could cover up other disciplines such as law and medicine.
- Sports interpreting shows significant similarities with community interpreting and is highly interactive.
- As a result of the universality of sports, sports terminology shows similarities in different languages, resulting in the use of loan words in the target culture.
- Its prescriptive characteristic is mostly observed in written language and sports texts should be translated accordingly.
- Sports news involves high level of aesthetic language use as well as creativity, and a sports Translator should transfer these features into the target language so as to accomplish functionality and the desired effect in the target culture.

Considering all these aforementioned characteristics, a sports Translator should render all the specific features of sports language. Moreover, being aware of the existence of both foreseeable and unexpected situations, he/she should acquire the necessary background information. Last but not least, in undergraduate programs, students should be given examples of such situations that sports Translators face in real life and should be taught ways on how to deal with these situations. Since Sports Translation is a new specific field in Translation Studies in itself, the future sports Translators will definitely require specialization in this field.

References

Altay, Ayfer. (2002). İngilizce ve Türkçe Hukuk Dillerinin Özellikleri: Karşılaştırmalı Bir Çalışma [Characteristics of English and Turkish Legal Languages: A Comparative Study]. Ankara: *Hacettepe Üniversitesi Edebiyat Fakültesi Dergisi* [*Hacettepe University Journal of Faculty of Letters*]. Vol: 19 No: 2. (pp. 13–32).

Bulut, Alev. (2011). Spor Dilinin Çevirisi: Çok Dramatik Bir Final Oldu [Translation of Sports.... Language: A Very Dramatic Final]. *Sözcükler.* 2011/1 No: 29. (pp. 64–66). Istanbul: Yazın Basın Yayın Matbaacılık Turizm Ltd. Şrk.

Kondo, Masaomi and Tebble, Helen. (1997). Intercultural Communication, Negotiation and Interpreting. *Conference Interpreting: Current Trends in Research Proceedings of the International Conference on Interpreting: What Do We Know and How?.* Yves Gambier, Daniel Gile, and Christopher Taylor (Eds.). (pp. 149–166). Amsterdam/Philadelphia: John Benjamins Publishing Company.

Linell, Per. (1997). Interpreting as Communication. *Conference Interpreting: Current Trends in Research Proceedings of the International Conference on Interpreting: What Do We Know and How?*, Yves Gambier, Daniel Gile, and Christopher Taylor (Eds.). (pp. 49–67). Amsterdam/Philadelphia: John Benjamins Publishing Company.

Munday, Jeremy. (2012). *Introducing Translation Studies: Theories and Applications.* (2nd ed.) London and New York: Routledge.

Pöchhacker, Franz. (2004). *Introducing Interpreting Studies.* London: Routledge.

Uyanık, Gözde Begüm. (2015). *Spor Çevirmenliğinde Durumların ve Görevlerin Tanımlanması [Defining the Contexts and Tasks in Sports Interpreting].* Master Thesis, İstanbul: İstanbul Üniversitesi [Istanbul University].

Vermeer, Hans J. (2004). Skopos and Commission in Translational Action. *The Translation Studies Reader.* Venuti, Lawrence (Ed.). (2nd ed.). (pp. 227–238.). New York and London: Routledge.

Vinay, Jean-Paul and Darbelnet, Jean. (1958/1995). *Comparative Stylistics of French and English: A Methodology for Translation.* Juan Sager and Marie-Jo Hamel (Tr.s and Eds.). Amsterdam and Philadelphia: John Benjamins.

Vinay, Jean-Paul and Darbelnet, Jean. (1958/1995). A Methodology for Translation. In *The Translation Studies Reader.* (2004). Venuti Lawrence (Ed.). (pp. 128–137). London and New York: Routledge.

Wadensjö, Cecilia. (2001). Community Interpreting. *Routledge Encyclopaedia of Translation Studies.* Baker, Mona (Ed.). (2nd ed.). (1st ed., 1998). (pp. 33–37). London and New York: Routledge.

Wang, Lidi and Zhang, Jie. (2011). Community Interpreting in China Since the Beijing Olympics 2008 – Moving Towards a New Olympic Discipline? *Modelling the Field of Community Interpreting: Questions in Methodology in Research and Training.* (pp. 263–279). Wien: LIT VERLAG GmbH and Co. KG.

Yücel, Tahsin. (2004). *Söylemlerin İçinden [Through Discourse].* (2nd ed.) İstanbul: Alkım.

Electronic Sources

American Bar:
http://www.americanbar.org/publications/tyl/topics/professional-development/what-is-sports-law-who-is-sports-lawyer.html.

FIBA:
http://www.fiba.com/

FIBA. (February 2014). *Official Basketball Rules 2014.* (Approved by FIBA Central Board). Barcelona. https://www.fiba.com/downloads/Rules/2014/Official_Basketball_Rules_2014_Y.pdf

FIBA U18 European Championship Men:
http://www.fibaeurope.com/compID_8aYeHlfuGF-mF5IqO8aFH1.season_2014.html

Hürriyet:
http://www.hurriyet.com.tr/telekom-kazanda-boguldu-10921949.

Prof. Mine Yazıcı Ph.D.

Istanbul University
Department of Translation Studies

Challenges and Suggestions in Medical Translator Training

Abstract: There is a great demand for medical translation in the face of a great number of patients admitted to the state or privately-run hospitals. The question is to what extent translator training programmes in Turkey can meet the demands of the foreign patients. Bearing this question in mind, this paper not only discusses the challenges in medical translation programmes within the framework of Turkish Higher Education System, but it also makes suggestions in consideration for the institutional constraints, linguistic differences, and learning habits of the trainees in the department of Translation Studies. After enlisting the challenges waiting for the medical translators, it discerns the translation procedures of medical professionals from those of medical translators. Within this framework, it discusses the limits and extent of specific field knowledge and knowledge of medical jargon from the perspective of medical translators. All in all, it discusses lexical, syntactic, and stylistic features of medical texts as well as the models and techniques in training prospective candidates of medical translation.

1. Introductory Remarks

Medical Translation is a subfield of not only technical communication, but it is also subsumed under the category of scientific translation. As a subfield of technical translation, it aims to transfer medical information as clearly and effectively as possible. Apart from this, as a subfield of scientific writing, it bears similar stylistic concerns as in literary translation in conveying and processing information in scientific texts (Byrne, 2012, pp. 2–3). Compared with the other fields of technical translation, there has been an increasing demand for interpreters and translators in the field of healthcare and medicine, especially when considered the huge number of refugees, migrants, and visitors of health tourism travelling all over the world. For example, the high number of international conferences held in Istanbul as one of the most visited cities in the world as well as the demand for publications in international journals of medicine in addition to those raising health awareness prove to what extent medical translation is important in academic environment nowadays.

Naturally, the huge amount of knowledge and the recent developments recorded in the field of medicine urge scholars to co-operate and maintain international correspondence with each other in the international arena. Since English becomes lingua franca in medical correspondence, it affects the direction of translation from Turkish or other languages into English. Namely, when the unilateral flow of knowledge from English to other languages has changed through the impact of globalization, the course of translation has become bidirectional; hence, it inevitably ends in mutual exchange of knowledge network. The reversal of direction from other languages including Turkish into English, in a way changes the definition of translation as a fact of target culture, since it becomes a fact of global culture. In this case, the linguistic and cultural problems concerning the source language foreground translation problems concerning the target culture. This is also the reason why scholars studying on medical translation focus on linguistic errors rather than translation problems, which originates from translators' subjective translation difficulties due to lack of linguistic, cultural, and transfer competences as well as due to lack of knowledge of documentation. This inevitably affects the ways of teaching medical translation and poses serious challenges at higher education. These challenges can be elaborated as challenges in medical translation and institutional challenges in medical training.

2. Challenges in Medical Translation

The challenges observed in medical translation can be classified into four groups. The *first challenge* is that Turkish medical language is full of loan words or terms transferred from Latin, Greek, French, and English. There appears to be three types of trends in Turkish medical language: those of French, English, and the purist movement.

The French trend adopted French medical terms or French terms in Turkish orthography until the 1980s. For example, the Turkish equivalent of French loan term 'systolique' is 'sistolik' in Turkish medical language. '<u>Chy</u>lorrhea' is another French loan word. Similarly, professionals have adopted French orthography and pronounced another word as 'şilore' [<u>sh</u>ilore] in Turkish orthography.

After the 1960s, there emerges the second stage, the age of English trend in Turkish medical language. In this period, due to the advances recorded in the field of medicine in the USA and England, medical terminology from Anglo-Saxon medical literature began to influence the international medical language. Accordingly, the professionals adapted English loan terms to Turkish orthography. For example, the prefix of the loan term 'chymus' is pronounced as 'kimus' [<u>kı</u>mus] as in British accent as opposed to 'şilore' [shilore] and 'chylorrhea' of the previous French examples (Ertem, 1998, pp. 13–36).

After the 1970s, there comes the last trend of the purist movement where linguists and scientists advocated 'purism' in scientific language, thereby suggesting neologisms or pure Turkish equivalents of foreign terms. As a result, a large number of terminological dictionaries were issued by the Turkish Language Association from the 1970s till the 1980s. For example, 'inflammation' is a loan word. It is still used and pronounced as 'inflamasyon', or 'enflamasyon' by professionals, yet the purists suggested 'yangı' in *Hekimlik Terimleri Kılavuzu* [Medical Terms Dictionary] (1978). It is still used in textbooks and academic language although the loan term 'inflamasyon' is also popular. On the contrary, the Turkish equivalent of 'otorhinolaryngology', a subfield of medicine, was 'otorinolaringoloji' in the past. However, in the course of time Turkish equivalent of the term 'oto+rhino+laryngo+ logy' is adopted as 'kulak burun boğaz hastalıkları bilimi'. It is understandable to native speakers and heralds a new trend in Turkish medicine 'to be understandable to native speakers' (Pilegaard, 1997, p. 170).

The *second challenge* to touch on Turkish medical language is the impact of western writing tradition on Turkish. Initially, even original textbooks read like translationese since medical professionals acquire professional knowledge by reading foreign resources so as to seize the recent knowledge produced in the west. Accordingly, their professional language develops in the third code[1] called 'translation'. Besides, medicine (and/or healthcare) is very complex and broad area of study. That is to say, it involves knowledge of specialties such as obstetrics, surgery, pediatrics, gynecology, orthopedics, psychiatry, and so forth.

The *third challenge* is related to the fact that medicine concerns both the public sector and private sector knowledge. Therefore, it requires the knowledge about not only specifics of genres alongside text-typology, but also the expectations of medical professionals as well as laymen from medical translation. The area of healthcare translation is very exhaustive and involves a large number of genres and subfields of study. Vicent Montald Rusereccio and Maria Gonzales Davies categorize the medical texts, in terms of the specifics of genres, as research genres and professional genres since nobody can expect a medical translator to be equipped with medical knowledge peculiar to specialists. While research genres involve research papers, reviews, clinical trials, protocols, meta analyses, doctoral dissertations,

1 'Third code' is used as a language peculiar to a group of people. Scientific jargon of a field of study can be assumed as third code since only professionals sharing the same field of study can maintain correspondence with each other.

scientific editorials, book reviews, and conference proceedings; professional genres cover clinical guidelines, informed consents, questionnaires, bulletins, annual reports, manuals, terminology glossaries, medical histories, maintenance guidelines, and so on. There are also educational and commercial genres. While the former involves patient information leaflets, fact sheets, presentations, TV documentaries script, textbooks, medical encyclopaedias and so on; the latter involves patents, contracts, packaging inserts, catalogues, and so forth (Rusereccio and Davies, 2007, pp. 30–33). This can be even extended to documents concerning health tourism.

Finally, the fact that medicine requires lifelong development of knowledge since new knowledge is replaced with the old one in a very short span of time constitutes the *fourth challenge.*

3. Institutional Challenges in Medical Training

In the light of these introductory remarks, there appears the second group of challenges: to what extent should medical translation be taught, especially when the complexity and the variety of fields of medicine and healthcare are taken into consideration? Similarly, the way how the medical or healthcare courses should be academically shaped in Turkey poses another question when the aforementioned constraints are taken into account. Amongst these constraints are the duration of courses, students' knowledge of technical writing, or technical translation, instructors' medical background or level of medical knowledge, and students' interest or familiarity with medical issues. In Istanbul University, for instance, the students take translation courses for specific fields in the 3rd and 4th grades, totally four hours a week. The two groups of courses are named as specific field knowledge and specific field translation. The trainers of the specific field knowledge courses are expected to be professionals or specialists; however, the professionals or specialists who taught the students from 2000 until 2014 are not currently allowed to teach specific field knowledge due to lack of financial resources of the university. Today only the academic staff of the university is allowed to teach specific field knowledge and translation courses. When the academic staff from the medical faculties are invited to give the courses on the specific field knowledge, there emerges two reservations: firstly, they are too busy with their students of the medical faculties; therefore, they are not able to teach the students of the department of translation studies.

Secondly, for the same reason they cannot get into contact with the instructors teaching medical translation. Apart from this, the academic staff of the department of translation studies instruct only for two hours to the students who have not taken technical writing courses yet since there are only writing courses on the basics of writing skills. This may seem to be the failure of scheduling, but in reality, this is related to the lack of space for courses. Accordingly, the number of courses has to be reduced as much as possible and technical writing should be taught within the coverage of specific field translation. As for the medical knowledge of the academic staff of the translation department, they cannot be expected to be equipped with as much medical knowledge as doctors or specialists performing the task. Nevertheless, despite all these challenges and constraints, instructors can at least do their best to set students off on path of lifelong learning.

Within this framework, the discerning features of medical translators and medical professionals who do translations can be elucidated so as to suggest instructors what strategy they should follow in the face of the aforementioned challenges and constraints.

4. Differences Between Medical Translators and Medical Professionals

Both medical translators and medical professionals can do medical translation. Nonetheless, it cannot be expected from medical professionals to allocate all their time to translation. Still, there is a common misconception amongst the trainees that medical translation is easy in comparison to other courses. Since medical language is full of foreign terms, the trainees suppose that medical translation is just an act of transferring terms from one language to another. However, if it were so easy, medical professionals would not need the assistance of medical translators in writing their scientific papers. Moreover, all the documents they need could be translated via machine translation. In other words, becoming a medical translator cannot be reduced to the literal transfer of knowledge from one language to another. First of all, it concerns 'health' and a small mistake in translation may even end in the death of a patient. Next, it requires being sensitive to linguistic and cultural issues since health organizations today aim to provide international services to patients or professionals all over the world.

In what ways medical translators differ from medical professionals doing translations can be examined in terms of the addressees and the mental procedures of medical translators and medical professionals.

4.1. The Difference in Addressees

The major difference between the medical translators and medical professionals is that medical translators address to a wider range of addressees. For this reason, they have to be aware of the reading habits peculiar to the target culture. Today the number of health-conscious readers has increased and medical knowledge does not concern only professionals. Accordingly, their mission should be to convey as much information as possible to readers at every educational background and social status and at every age. Medical professionals, however, mainly concern about sharing information with their colleagues and they share their knowledge in their own scientific jargon. In a way, they preserve the canonicity of their field of study by developing and maintaining correspondence in their own jargon. However, today especially from the perspective of patients' rights, World Health Organization (WHO) proposes clarity in language as it is declared in the following statement[2]:

> Each of these suggests different professional obligations of the physician toward the patient. For instance, in the paternalistic model, the best interests of the patient as judged by the clinical expert are valued above the provision of comprehensive medical information and decision-making power to the patient. The informative model, by contrast, sees the patient as a consumer who is in the best position to judge what is in her own interest, and thus views the doctor as chiefly a provider of information.

At this point, medical translators' knowledge of the reader profile as well as that of linguistic and stylistic, and the specifics of genres are helpful in informing the patients of their rights, and of their diseases and ailments by using a clear and understandable language.

4.2. The Difference in the Mental Procedures of Medical Translators and Medical Professionals

Another prominent difference between medical translators and medical professionals may arise from comprehending and processing the source text from different poles of the total knowledge base. In other words, translators' knowledge base, which is composed of linguistic knowledge and extra-linguistic, or worldly knowledge, is different from medical professionals' total knowledge base in starting translation, particularly at the beginning of the translation process. Medical professionals' knowledge base is composed of the specific field knowledge and the knowledge acquired from the histories of patients, or their case studies. In other words, the specific field knowledge is related to medical translators' performances on patients. Accordingly, it can be asserted that the total knowledge

2 This information was retrieved from the website of World Health Organization (WHO) on 7.5.2016.

base of doctors is constructive, whereas medical translators' knowledge is based on the documents they have translated.

Therefore, it is passive knowledge acquired from the texts. Apart from this, medical professionals can relate the terms to those they have already acquired from the specific field knowledge, or from their worldly experience they have gained from their performances on the patients. Accordingly, terminological knowledge and specific field knowledge play an active role in solving translation problems and understanding the source text. Since their knowledge of terminology is wider and more accurate, they can integrate the new information presented in the source text into their specific field knowledge base. In other words, a medical professional can refer to his/her specific field knowledge in such a way as to expand the network of relations between the terms in the process of translation, while medical translators expand their terminological or encyclopaedic knowledge without any reference to their specific field or professional knowledge as translators (Yazıcı, 2011, pp. 89–98). They acquire specific knowledge in the course of translation, which only serves to expand their total knowledge base. That is the reason why terms are one of the most difficult barriers in translation because the knowledge of terms is not so accurate for them as for medical professionals. Nevertheless, they can refer to their linguistic knowledge and research skills to overcome terminological difficulties. Accordingly, spotting relations between the linguistic data will help the medical translators to understand the terms and expand their terminological knowledge. The following example[3] may illuminate the translator's comprehension of the source text and the way they acquire from the text at first sight:

Example 1

> BCG VACCINE
>
> Introduction
>
> BCG, or bacille Calmette-Guerin, is a vaccine for tuberculosis (TB) disease. Many foreign-born persons have been BCG-vaccinated. BCG is used in many countries with a high prevalence of TB to prevent childhood tuberculous meningitis and miliary [sic][4] disease. However, BCG is not generally recommended for use in the United States because of the low risk of infection with *Mycobacterium tuberculosis*, the variable effectiveness of the vaccine against adult pulmonary TB, and the vaccine's potential interference with

3 The example was retrieved from the website of the U.S. Department of Health and Human Services on 7.5.2016.

4 [sic] indicates that although the word looks odd or wrong, it is intended to be written like that or the original writer wrote it like that.

tuberculin skin test reactivity. The BCG vaccine should be considered only for very select persons who meet specific criteria and in consultation with a TB expert.

Analysis of Source Text
BCG
Something called BCG is a vaccine for tuberculosis. Its expansion is Bacille Calmette-Guarin. The eponym BCG is obviously derived from the name of two scientists who invented vaccine against tuberculosis. It is used in countries where Tuberculosis (or its acronym (TB) is widespread. However, the vaccine is not generally recommended in the USA because of the low risk of bacterium called Mycobacterium tuberculosis. Second, it may not be so much effective in adults suffering from Tuberculosis as in children. That is to say, its effectiveness may change in adults. Furthermore, it interferes with skin test and cause reactions. For this reason, only certain people meeting certain criteria, or only those under the consultancy of a specialist patients are recommended to be vaccinated.

In the beginning the translator may not relate the newly acquired knowledge to his/her total knowledge base as medical professionals do. Yet, as he/she proceeds, the newly acquired knowledge is added to the knowledge base. That is why Daniel Gile (1995, p. 110) claims, "The knowledge base is highly dynamic in its content, at least as far as the information relevant to the ongoing translation is concerned." Briefly, medical translators also acquire new knowledge in the course of translation by establishing correlations between linguistic units. In the beginning the knowledge they acquire from the source text is isolated. However, it becomes functional as they proceed; in other words, the professionalism of medical translators is based on their knowledge as an intercultural intermediary. As a competent reader and sender of the text they can reformulate the target text better than the medical professionals can do. For example, the following abstract issued in the journal of İstanbul Medical University[5] is translated from Turkish into English. The clear and plain style as opposed to the complex sentences proves the concern of medical translators for sharing as much knowledge as possible with the readers as opposed to the past concern for ornate style and highly academic language in the past:

Example 2

D VİTAMİNİ EKSİKLİĞİNDE RİSKLİ GRUPLARDAN BİRİ: SAĞLIK ÇALIŞANLARI
Amaç: D vitamini vücudumuzda kalsiyum, fosfor metabolizması ve iskelet sistemi üzerine etkili olan, önemli miktarı güneş ışınları ile deride sentezlenen ve diyetle alınabilen bir hormondur. Güneşe maruziyetin yetersiz olduğu ortamlarda çalışmak D vitamini eksikliğine neden olabilir. Yetersiz güneş alan binalarda ve gece nöbetlerinde

5 Varkal M A, Yıldız İsmail, et al., retrieved from www.journals.istanbul.edu.tr. on 7.05.2016.

> çalışmaları nedeniyle sağlık çalışanları önemli risk grubunu oluşturur. Çalışmada, sağlık çalışanlarının D vitamini düzeyleri üzerine, çalışma ortamlarının ve zamanlarının etkilerini belirlemeyi amaçladık.
>
> ONE OF THE RISK GROUPS FOR VITAMIN D DEFICIENCY: HEALTH WORKERS Objective: Vitamin D is an essential hormone in calcium, phosphor and bone metabolism. Skin synthesizes great amount of vitamin D in sun light, [sic] it can also be taken by diet. Vitamin D deficiency may occur in settings with low sunlight exposure. Medical personal [sic] are under risk for vitamin D deficiency, especially who works [sic] at night shifts and in low sun-exposed buildings. Our aim was to determine vitamin D levels of medical personal [sic] and evaluate [sic] effect of working time and place on vitamin D levels.

As seen here, the transfer of knowledge foregrounds the direct transfer of signs. The sentences are divided into simple sentences. The translator adopts *simplification* which is universal in consideration for the terminological density and compact nature of medical language. Accordingly, the translator has a recourse to *condensation strategy* in translating medical texts by observing the syntactic features of medical English. One can relate these translator's decisions to the *referential* and *conative function* of medical texts. In other words, the translator considers the dual function of the medical texts as informing or depicting a situation, and inviting the readers to perform. However, in spite of these functions, linguistic features of medical text generally pose challenges in terms of the high number of premodifications, nominalizations, passive constructions, and reduced relative clauses. In this simple example given above, it is observed that the translator splits the first sentence, and has a recourse to ellipsis and nominalization so that the text becomes more comprehensible and legible for the addressees:

Split of Complex Sentences

> "D vitamini vücudumuzda kalsiyum, fosfor metabolizması ve iskelet sistemi üzerine etkili olan, önemli miktarı güneş ışınları ile deride sentezlenen ve diyetle alınabilen bir hormondur."
> "Vitamin D is an essential hormone in calcium, phosphor and bone metabolism. Skin synthesizes great amount of vitamin D in sun light, [sic] it can also be taken by diet."

In the example written in the source language (Turkish), there is only one statement whose meaning is transferred with two sentences (even three sentences if a semicolon is used instead of a comma.)

Ellipsis 1

> "Güneşe maruziyetin yetersiz olduğu ortamlarda <u>çalışmak</u> D vitamini eksikliğine neden olabilir."
>
> "Vitamin D deficiency may occur in settings with low sunlight exposure."

The English equivalent for the word 'çalışmak' [to work] is not used in the sentence.

Ellipsis 2

> "Yetersiz güneş alan binalarda ve gece nöbetlerinde çalışmaları nedeniyle sağlık çalışanları önemli risk grubunu oluşturur."

> "Medical personal are [sic] under risk for vitamin D deficiency, [sic] especially who works at night shifts and in low sun-exposed buildings."

The Turkish word 'nedeniyle' which is employed to show the cause of vitamin D deficiency is not transferred to the sentence written in target language (English).

Nominalization

> "Çalışmada, sağlık çalışanlarının D vitamini düzeyleri üzerine, çalışma ortamlarının ve zamanlarının etkilerini belirlemeyi amaçladık."

> "Our aim was to determine vitamin D levels of medical personal and evaluate effect of working time and place on vitamin D levels."

The verb 'amaçla-dık' [we aimed…] is changed into the nominal usage of 'our aim'.

As seen in these examples, *brevity* is one of the main features of medical translation from Turkish into English. However, brevity does not mean medical translation is easy. On the contrary, *condensity* and brevity of medical texts require high level of specialized background knowledge. Even if a medical translator knows the features of medical language in abstracts, s/he needs to get into contact with a medical professional so as to avoid misunderstandings in the target text. Briefly, medical translation is a team work. It achieves its end only if medical translators and professionals cooperate with each other. Particularly in medical texts of high specialization, the consultancy to medical professionals will optimize the medical translator's task in terms of comprehensibility, legibility, and retrieval of correct information. For example, Rand Health is a non-profit institution of research and analysis and produces many practical surveys and public documents. On their internet page of the institution, one can easily reach the following translation guidelines for surveyors[6]:

- Translator(s) is briefed on socio-demographic characteristics of target population, mode of administration of the survey instrument, and where the survey will be administered;
- Provide translator with specific instructions about the reading level he/she should be aiming for in the translation and whether he/she should use language

6 The guidelines were retrieved from the website of Rand Organization on 7.5.2016.

that is going to be widely understood by a variety of speakers of the target language or whether the translation should reflect language usage by speakers from a particular region or country;

- Translator(s) reviews original language survey instrument before translation to identify items, terms or concepts that are difficult to translate;
- Translator(s) meets with survey user to discuss problem items, terms or concepts and to obtain additional clarification or information on goal or intent of the English language item, term or concept;
- Translation into target-language by a professional translator who is a native speaker of the target language (preferably a certified translator);
- Back-translation from target-language into English by professional or certified translators;
- Review of translation by bilingual reviewers (or other professional translators);
- Review of original English-language instrument and back-translation;
- Resolve discrepancies or problems in the translation by a committee that includes the translator, back-translator and reviewers (this may require one or more phone or in-person meetings).

It is interesting that the guideline addresses directly to the surveyors. Accordingly, the surveyor is positioned as a commissioner, who is entitled to draw up the brief of translation task. Furthermore, one can easily observe that a medical translator is expected to act not as an expert, but as an intermediary in transferring information under the guidance of medical surveyor. The question is today professional translators are expected to be equipped with most of the knowledge mentioned above in the guidelines. They are aware of the text specific features of questionnaires and cultural issues concerning the targetees. The dilemma is that the guidelines are arranged not for the medical translators, but for the surveyors although medical translators as experts in the field of translation are responsible for the translation task.

This way of approach is just opposed to Skopos Theory, which places the translator in the position of bi-cultural expert (Vermeer, 1996, p. 35). Theoretically and professionally, two inferences may be drawn here: first, the translators are still expected to act invisibly; next, there is not enough correspondence between medical translators and medical professionals. This is why the above-mentioned guideline is arranged for medical professionals to act as both commissioners and experts as representatives of canonized field of study. It aims to provide medical professionals with cultural and linguistic awareness to obtain sound data from the survey. Undoubtedly, the cooperation between the medical professionals and medical translators will consolidate the findings of the survey. On the other hand, studying such guidelines will yield the trainers, clues in shaping medical translation courses in higher education. Nonetheless, before discussing medical training, it would be much better to see what challenges are waiting for the translators in Turkish medical language.

5. Text-types in Turkish and English Medical Language

As a subfield of technical writing, text types in medicine are categorized according to the terminological density. Peter Newmark's classification of text types as academic, professional, and popular also applies to text types in medical translation. Accordingly, texts dense in Latin and Greek loan words are accepted as 'academic' and they are associated with academic papers; whereas semi-formal language used by medical professionals is called 'Professional'. Finally, popular medical language used by layman is called 'popular' text types (Newmark, 1988, p. 153). Similarly, Asalet Erten, who is the only author of book in Turkish in the field of medical training, adopts Newmark's classification; however, she prefers James Hudson's (1978) terms in classification of medical texts since the terms N. K. Hudson used is easier for Turkish trainees to classify text-types (cited by Erten 2003(a), p. 107). Accordingly, she uses high-level technical language, intermediate-level technical language, low-level technical language (Erten, 2003(a), pp. 105–107). Nevertheless, it is difficult to make such a definite classification in Turkish since the borders between academic professional and popular texts are fuzzy, and there is a huge gap between layman's medical language and professional language. The following example may illuminate this:

Example 3

Perkütan transhepatik kolanjiyografinin evaluasyonunda primer (inisiyal) kriter duktusların opasifikasyonudur. Parametrelerden kontur irregülaritesi, striktürler, inhomojenite ve defektler sayılabilir. Klinik diagnoz, radyolojik diagnostik metodlarla konfirme edilmelidir.[7]

Back Translation

In the evaluation of Percutaneous transhepatic cholangiography the criterion is opacification of primary (initial) ducts. Irregularity of contour of the parameters, strictures, inhomogeneity and defects are counted. Clinical diagnosis should be confirmed by radiological diagnostic methods[8].

---

7 Akan H., retrieved from www.dirjournal.org on 21.4.2017.

8 'Tıpta Türkçe Kullanımı' [Using Turkish in Medicine], retrieved from the website of patoloji.gen.tr. on 8.5.2016.

Example 4

> Transözofageal ultrasonografinin rezolusyonundaki artış sayesinde, mediastinal lenf nodlarınızın ne kadar atake olduğunu detaylı olarak vizualize edebileceğimizi söyleyebilirim.
>
> ***Back Translation***
> With the increase in the resolution of the transesophageal ultrasonography, I can tell you in detail to what extent your mediastinal lymph nodes are attacked.

Turkish is an agglutinative language and suffixes determine the function of the words in the sentence. In both examples only the suffixes, for instance, 'ultrasonografinin', 'rezolusyonundaki', 'kolanjiyografinin' are in Turkish, but the rest of the both texts are full of loan medical terms which undergo orthographic shifts in Turkish. It is for this reason that students claim medical translation is easy since it is just an act of direct transfer of terms and the insertion of suffixes to them. In this case, the texts address much more to English professionals than to Turkish layman. In other words, Turkish patients would not understand anything when the medical professionals use the above-mentioned language in explaining the medical procedures applied on him/her. However, the layman should understand the procedures as mentioned in the statement from the Universal Declaration of Human Rights formalized in 1948[9]: "The patient as a consumer who is in the best position to judge what is in her own interest, and thus views the doctor as chiefly a provider of information."

This may be owing to the fact that medical knowledge is acquired through foreign resources and Turkish medical language develops in the third code. Even most of the original textbook or academic papers in Turkish are written in the third code if the authors lack linguistic competency. Therefore, the main question concerning medical training could be handled by providing the trainees with medical linguistic awareness and the classical IMRAD (Introduction, Material and Methods, Results, and Discussion) structure in translating medical research papers, and helping them avoid lexical and sentential misuses or linguistic errors in translation (Pilegaard, 1997, p. 177).

At this stage starting from the popular texts may be a solution to familiarize the students with the medical language and the problems encountered in medical translations. In this way, they gain awareness why they should learn certain amount of medical terminology and medical language. It would enhance their encyclopaedic knowledge, which is an essential part of their comprehension.

---

9 Universal Declaration of Human Rights, retrieved from the website of World Health Organization (WHO) on 7.5.2016.

Asalet Erten's translation of Vus'at O. Bener's short story of 'Brucella' (1998) may be a good example in starting medical translation courses since the problems in medical translation in a short story may arouse curiosity in trainees and explains why they should be equipped with a certain amount of medical knowledge. In other words, a trainer can achieve the first stage of learning called 'problematizing'. This example is full of medical terminology, and the translator, Asalet Erten, who is a trainer of medical translation, exemplifies in what way she has a recourse to her knowledge of medical translation and overcome terminological problems:

Example 5

Source Text

Aldığım ilaçlar hayli kabarık. Belli saatlerde veriyor hemşireler ilaçları. Dört miligramlık Cardura (tansiyon için), bir adet Drisentin **(damar açıcı),** bir adet Rytmonorm (taşikardimi düzenleyici) bir adet Trental-400 **(kalbi korumaya yarıyormuş)** bir adet Hydrea (kandaki aşırı yüksek trombosit sayısını azaltmaya yarayan) (cited by Erten, 2003(b), p. 31)

Target Text

I am overloaded with lots of medicine. The nurses give the medication at certain hours. Four milligram Cardura (for blood pressure), one tablet of Drisentin **(for vasodilatation**), one tablet of Rytmonorm (for controlling tach-cardia) one tablet of Trental-400 **(for cardiac prophylaxy),** one tablet of Hydrea (for decreasing the extremely high trombosit count in the blood) (translated by Erten, 2003(b), p. 31).

Although the author uses colloquial language ('damar', 'damar açıcı', 'kalbi korumaya yarayıcı') in explaining the disease, except for the brands of drugs; the translator uses medical terminology in consideration for the English readers. She uses 'vasodilatation' instead of saying 'damar açıcı' [opening veins] or 'cardiac prophylaxy' instead of 'kalbi korumaya yarıyormuş' [protecting heart]. She is not a medical professional, but as a medical translator and trainer she refers to her knowledge of medical terminology in literary translation. This, on the one hand, indicates that medical translation is not peculiar only to medical texts; on the other hand, today is the Age of Information and the readers demand to know and share knowledge in every field of study. Starting with such examples will gain trainees awareness why they should take medical translation classes. If such an example is planned to be used for medical translation courses, the students may be asked to read the whole story in Turkish and to spot passages concerning medical terminology and to translate them in academic, professional, or layman's language. After that, their decisions and the resources they referred to in overcoming translation problems may be discussed. Finally, the trainer may ask to compare the English version of the story with their own translations so as to see why they require knowledge of basic medical terminology.

Trainers may expand the examples by giving such texts concerning daily life as case studies, histories of patients, and so forth. Starting from the popular texts they are familiar with and ending the courses with academic texts will help trainees to systematize their knowledge. Doing intralingual exercises, or asking them to edit, or forming a text explaining the case of the patient will also help them to discern medical jargon from standard English.

Example 6

> 42 years old female patient **admitted** to Istanbul Unıversity Internal Medıcıne Endocrinology Department **for a 3 years of history of hypertension**, headache, tachcardia attacks, and increasing symptoms like these during bending forward. But later she **referred** to Gynaecology Department. After she was treated, the hospital **discharged** her.
>
> **A woman aged** 42 years **was** admitted to Istanbul University, Internal Medicine Endocrinology Department, with a 3-year history of hypertension, headache, tachycardia attacks. These symptoms **worsened when she bent forward. Later, she was referred to** Gynaecology Department. After she was treated, she was discharged from the hospital.

When the students were first asked to edit the first excerpt, they could not do it. There are two reasons for doing this exercise: Firstly, to teach them what words ('refer', 'admit', 'discharge', and so on) are used in admittance to a hospital and secondly, to show why they should learn medical jargon. In this example, students' lack of medical jargon is one of the main problems in editing the text. To put it in a different way, adult learning is different from children's learning. They want to understand why they learn and identify what they should learn themselves. In this case, one can infer that trainees' problem cannot be limited to the lack of medical knowledge, but also to the trainer's strategy in teaching medical translation. At this stage, while textbooks for Specific or Professional Purposes are helpful in teaching English medical jargon, the trainers' awareness of translation-oriented analysis will also help students to gain not only the basics of medical knowledge, but also help them to question medical translation in two directions; namely, from English to Turkish and in reverse direction.

6. Concluding Remarks and Suggestions

Trainers should be aware of the expectations of targetees in medical translation. For example, in Turkey, most of the specialists in the field of medicine know English as a foreign language and can translate their own original texts into English. In this case, they only require optimization of their texts in English. At this stage, medical translators mostly act as editors and make linguistic revisions on the text. Meanwhile, the main demand from medical translators is to translate from

Turkish into English. Accordingly, as trainers we should always keep Paul Grice's Cooperation principle in mind in training even it is in the specific field. In other words, students should be told that translators are responsible for maintaining communication between the involved parties.

For this reason, translators should always consider Grice's Maxims of quality, quantity, relevance and manner, which means avoiding ambiguity and being brief and relevant. Grice's Maxims apply to medical translation since syntactic features and linguistic usages are brief and simple in medical translation. For instance, medical texts are full of premodifications, nominalizations, passive or impersonal structures as well as defining and non-defining clauses, or 'reduction of relative clauses'. For example, premodifications such as 'modified radical mastectomy' or 'newly formed lymphocytes' contribute to the brevity maxim. Similarly, the high number of nominalizations as in the following example show to what extent words can be economized:

> When we study cell structure we are limited by the techniques available.
>
> The study of cell structure is limited by the techniques available.

Meanwhile trainers can stress on the extraordinary usages in medical language. For example, in academic or scientific writing it is not proper to change (reduce) non-defining clauses into a short form or to omit relative pronoun 'which'. However, in medical writing one can omit it as in the following example:

> Glycogen, derived from glucose, is stored for future use.

By this way, trainees' attention can be drawn to the maxims of brevity and quantity in medical translation. This also means trainers should stress on punctuation in medical texts. Medical professionals may make mistakes in punctuation since language is not their field of study; however, medical translators should consider punctuation marks. They can be asked to search for similar examples in parallel texts in Turkish.

As for specific field knowledge, instructors are not expected to specialize in medicine. However, they can teach anatomy or anatomical positions of the body and the functions of the organs. At this stage, medical textbooks in Turkish and English will help students compare the same knowledge in both languages. Learning is constructive. Following this, they can be assigned to search for the texts concerning the related diseases of these organs as well as dialogues between doctors and patients on the internet pages of the medical sites.

Students may also be asked to go to the university or private hospitals to search for the documents concerning foreign patients and ask them the procedures necessary for the admittance, laboratory tests, epicrisis, or discharge reports as well as health insurances of foreign patients. In this way, they get acquainted with institutional organization, administration, and governance of hospitals, and gain awareness of the importance of documentation in medical translation.

As for teaching terminology, students cannot be expected to memorize all the Greek and Latin roots, prefixes, and suffixes. As mentioned before adult learning is different from children learning. The basic or frequently used roots, prefixes, and suffixes may be taught by giving such exercises refining their reasoning abilities. For example, instead of asking directly to memorize the meanings of such prefixes as 'intra-', 'inter-', or 'para-', they may be asked to paraphrase the terms in English using the medical terminology with the above-mentioned Greek or Latin suffixes or prefixes as in the following examples:

Fluid outside the cells ⟶ Extracellular fluid

Membrane around the tooth ⟶ Periodontal membrane

Such exercises will also help them to develop their transfer competence. Apart from this, drawing their attention on the misuses of Greek and Latin roots in Turkish will refine the quality of translations; for example, 'data' is plural in Latin; however, professionals generally add Turkish plural suffix 'lar' to the word and translate it as 'datalar' although the singular form of it is 'datum'. Similarly, 'emboli' is plural and the singular form is 'embolus'. These may seem simple errors, but in medical translation simple errors may end in wrong findings.

Undoubtedly, translating eponyms and acronyms is worth stating. For example, as eponyms, 'Parkinson's disease' becomes 'Parkinson Hastalığı' in Turkish or 'Behçet Hastalığı' becomes 'Behcet's disease' in English. Similarly, some acronyms are preserved in its original form even though some may have Turkish equivalent. For example, DIRV (Double-Inlet Right Ventricle), AAA (American Association of Anatomists), Ca (Cancer Antigen), and so on.

Teaching medical translation is not an easy task under institutional, spatial, and temporal constraints as it is aforementioned. One problem is that there are not technical writing and translation classes in most of the Turkish universities. Accordingly, teaching medical translation becomes more difficult since the trainees try to learn medical writing without any knowledge of technical writing.

Resources of medical translation in English, on the other hand, fall behind expectations in distant languages and cultures even if they help the instructors to draw up the general framework of medical translation classes. The trainers or instructors of medical translation classes should know the profile, educational background, and the requirements of Turkish students in shaping medical translation courses. All in all, in spite of all these constraints and challenges the instructors can set students off on the path of developing medical translation skills if they believe in constructive learning and if they can teach the logics of medical translation.

References

Byrne, Jody. (2012). *Scientific and Technical Translation Explained.* A Nuts and Bolts Guide for Beginners. New York: St. Jerome.

Ertem, Güngör. (1998). *Kardiyoloji Terimleri Sözlüğü* [Cardiological Terms Dictionary]. İstanbul: Türk Kardiyoloji Derneği [Turkish Society of Cardiology].

Erten, Asalet. (2003) (a). *Tıp Terminolojisi ve Tıp Metinleri Çevirisi* [Medical Terminology and Translation of Medical Texts]. Ankara: Seçkin.

Erten, Asalet. (2003) (b). Vus'at O. Bener'in Brucella Başlıklı Öyküsünün Çevirileri [Translation of the Short Story Brucella by Vus'at O. Bener]. *Çeviribilim Uygulamaları Dergisi.* Sayı/Number:13 (ss./pp. 25–35). Ankara: Hacettepe Üniversitesi Yayınları [Hacettepe University Publications].

Gile, Daniel. (1995). *Basic Concepts and Models for Interpreter and Translator Training.* Amsterdam/Philadelphia: John Benjamins.

Grice, Paul. (1989). *Studies in the Way of Words.* Cambridge, Massachusetts and London: Harvard University Press.

Hekimlik Terimleri Kılavuzu [Medical Terms Dictionary]. (1978). Ankara: Türk Dil Kurumu Yayınları [Turkish Language Association Publications].

Newmark, Peter. (1988). *A Textbook of Translation.* New York: Prentice Hall.

Pilegaard, Morten. (1997). Translation of Medical Research Articles in Anna Trasborg (Ed.) *Text Typology and Translation* (pp. 159–184). Amsterdam: John Benjamins.

Resureccio, Vincent Montalt and Davies, Maria Gonzales. (2007). *Medical Translation Step by Step.* Manchester: St. Jerome.

Vermeer, Hans. (1996). *A Skopos Theory of Translation.* Heidelberg: Text and Context.

Yazıcı, Mine. (2011). *Çeviribilimde Araştırma [Research in Translation Studies]*. İstanbul: Multilingual.

Electronic Sources

Akan, Hüseyin. (2003). Tıpça Üzerine Türkçe Düşünceler [Turkish Thoughts on Medical Language]. *Tanısal ve Girişimsel Radyoloji [Diagnostic and Interventional Radiology Journal].* 9:131–134
http://www.dirjournal.org/sayilar/49/buyuk/pdf_TGR_3661.pdf

Rand Organization
http://www.rand.org/health/surveys_tools/ about_translations.html

Tıpta Türkçe Kullanımı [Using Turkish in Medicine]
http://patoloji.gen.tr/

The U.S. Department of Health and Human Services (Tuberculosis)
http://www.vaccines.gov/diseases/tb/index.html

Varkal, Muhammet Ali, Yıldız İsmail et al. (2015), *İstanbul Tıp Fakültesi Dergisi [Journal of Istanbul Faculty of Medicine]* Cilt [Volume]: 78 • Sayı [Number]: 2 • Yıl [Year]: 2015
www.journals.istanbul.edu.tr/iuitfd/article/view/5000127452

World Health Organization
http://www.who.int/genomics/public/patientrights/en/